THE UPTAKE AND STORAGE OF NORADRENALINE IN SYMPATHETIC NERVES

THE UPTAKE AND STORAGE OF NORADRENALINE IN SYMPATHETIC NERVES

BY

LESLIE L. IVERSEN

Fellow of Trinity College, Cambridge

CAMBRIDGE
AT THE UNIVERSITY PRESS
1967

Published by the Syndics of the Cambridge University Press
Bentley House, 200 Euston Road, London, N.W. 1
American Branch: 32 East 57th Street, New York, N.Y. 10022

Library of Congress Catalogue Card Number: 67-12318

Printed in Great Britain
at the University Printing House, Cambridge
(Brooke Crutchley, University Printer)

CONTENTS

LIST OF PLATES

LIST OF TEXT-FIGURES

PREFACE

Since the publication of the comprehensive review on the catecholamine hormone and neurotransmitter, noradrenaline, by Professor von Euler ten years ago, there has been an almost explosive increase in the number of research studies devoted to this substance. Figure 1,

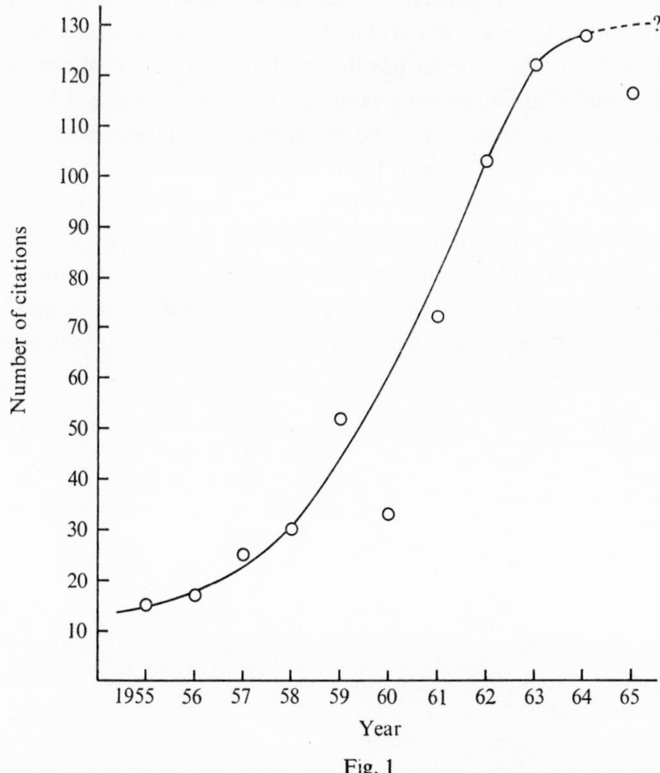

Fig. 1

which indicates the number of publications quoted in this work for each year since 1954, illustrates that at this time the field of catecholamine research appears to be in a phase of intense activity. Such a stage in the development of a subject might hardly seem an appropriate time for a monograph, since the author may be almost certain that much of his writing will be rendered incomplete or obsolete very rapidly.

Nevertheless, fools will always rush in where angels fear to tread, so that I gratefully accepted the kind invitation of the Cambridge University Press to extend and rewrite my Ph.D. thesis for publication as a monograph. This work is written and published, however, in the full knowledge that it can be regarded only as a progress report on a field of research which is rapidly growing, diversifying and still in a state of turmoil. The literature cited in this work includes published material up to the end of 1965, together with certain articles pending publication at that time. I am very grateful to the many authors who have been kind enough to allow me to use illustrations from their published work. I would also like to express my gratitude to Professors L. G. Whitby and A. S. V. Burgen, who supervised my research for the Ph.D. degree between 1961 and 1964. Without their constant advice, inspiration and encouragement during that period I certainly would not have been in a position to write this book. To my wife especial thanks, since without her patient long hours of work in compiling bibliographies this work might still not have reached the publisher's office. The monograph was prepared during my tenure of a Harkness Fellowship in the United States.

Boston, 1966 L. IVERSEN

1

INTRODUCTION

While fundamental developments have been made in recent years in studies of the structural and physiological properties of synapses, the immense biochemical complexity of the synapse is only now becoming apparent. The monographs of Eccles (1964) and McLennan (1963) are excellent and comprehensive accounts of recent developments in the study of the physiological, anatomical and biochemical properties of synapses. The present work is devoted to a review and discussion of recent developments in studies of adrenergic synapses, primarily those between the terminals of postganglionic sympathetic neurones and peripheral effector tissues. Although the author's bias is biochemical, it is clear that in the study of synapses no distinct lines can be drawn between scientific disciplines. The synapse is an extraordinarily complex structural and functional unit. A detailed knowledge of its anatomy and physiology are necessary for an understanding of the complex chemical functions which are carried out in this unit. Nowhere is it more true that studies of biochemical function cannot ignore the structural integrity and complexity of the tissue. While classical biochemistry made great strides in the elucidation of fundamental biochemical mechanisms in tissue homogenates in which cellular structure was destroyed, modern biochemists are becoming increasingly aware of the importance of cellular structure in regulating and determining the pattern of biochemical activity of intact tissues. This might be said to be the main thesis put forward in this review, that cellular structure can be equally as important as enzymic machinery in determining the chemical functions of an intact cell. Indeed, in the case of noradrenaline it is proposed that the transport of this material from one tissue compartment to another, without any enzymic degradation, can entirely account for the physiological inactivation of this amine after its release from adrenergic nerve terminals.

Unfortunately neurochemistry is as yet ill-equipped to deal with the problems of transmitter chemistry in individual synapses. While neuroanatomy and neurophysiology have entered a period of intellectual sophistication, with powerful micro-techniques for studying the detailed anatomical and electrophysiological properties of individual neurones, neurochemistry is still limited by the relatively crude and insensitive chemical tools available. The recent developments in chemical studies of adrenergic synapses represent a timely and fruitful beginning in the application of modern biochemical techniques to a study of the properties of synapses and the interaction of drugs with these structures. The present monograph presents a highly personal and biased review of these developments; for comprehensive reviews of much of the material presented here the reader is referred to the sources listed at the end of this chapter.

The study of noradrenaline and its function as the adrenergic transmitter substance has advanced rapidly in the last decade, and such studies cover an increasingly broad front of research activity. For example, the development of a sensitive and specific histochemical technique for visualizing catecholamines in tissue sections has led to great increases in our knowledge of the morphology and detailed anatomy of adrenergic neurones. It is now clear, for instance, that in the peripheral sympathetic nervous system the classical concepts concerning the morphological distribution of postganglionic nerve cell bodies in sympathetic ganglia are not always correct. In certain tissues, such as the vas deferens and accessory male genital glands, the postganglionic nerve cell bodies are in close proximity to the innervated tissue. In the brain, histochemical studies have revealed that noradrenaline, dopamine and serotonin are exclusively localized in specific systems of neurones—thus lending strong support to the view that these substances may serve as transmitters at synapses in the central nervous system. If this is eventually shown to be the case, the term 'adrenergic' may be replaced by the terms 'noradrenergic' and 'dopaminergic'; meanwhile the term 'adrenergic' is commonly used to describe chemical transmission at peripheral sympathetic nerve terminals, although it is understood that this should strictly be 'noradrenergic' since noradrenaline is now known to be the sole transmitter at such synapses.

Another fascinating line of research in the field of adrenergic mechanisms has been the discovery that a protein growth factor can profoundly influence the development of the peripheral sympathetic nervous system in young animals. The discovery of the 'sympathetic nerve growth factor' and the development of an antiserum to this protein, which can partially destroy the developing sympathetic nervous system, are fundamental advances which may have far-reaching consequences for studies of the development of the nervous system, and the formation of specific neuronal connections (Levi-Montalcini, 1964). In this monograph the term 'immunosympathectomized' will be used to describe adult animals in which the development of the sympathetic nervous system has been impaired by the administration of nerve growth factor antiserum in the first few days after birth.

BIBLIOGRAPHY

Monographs

Euler, U. S. von (1956). *Noradrenaline*. Springfield, Illinois: Charles C. Thomas.

Eccles, J. C. (1964). *The Physiology of Synapses*. Berlin: Springer-Verlag.

De Robertis, E. (1965). *Histophysiology of Synapses and Neurosecretion*. Oxford: Pergamon Press.

McLennan, H. (1963). *Synaptic Transmission*. Philadelphia: W. B. Saunders and Co.

Stjärne, L. (1964). Studies of catecholamine uptake, storage and release mechanisms. *Acta physiol. scand.* **62**, Suppl. 228.

Recent symposia

Adrenergic mechanisms. Ciba Fdn Symp. London: Churchill. 1960.

First International Catecholamine Symposium. *Pharmac. Rev.* **11** (1959).

Second International Catecholamine Symposium. *Pharmac. Rev.* **18** (1966).

Pharmacology of Cholinergic and Adrenergic Transmission. Second International Pharmacology Meeting, Prague, 1963. Oxford: Pergamon Press. 1965.

Mechanisms of Release of Biogenic Amines. Oxford: Pergamon Press. 1966.

Review articles

Axelrod, J. (1965). The metabolism, storage and release of catecholamines. *Recent Prog. Horm. Res.* **21**, 597–622.

Brodie, B. B. & Beaven, M. A. (1963). Neurochemical transducer systems. *Med. exp.* **8**, 320–351.

Carlsson, A. (1965). Drugs which block the storage of 5-hydroxytryptamine and related amines. In *Handb. exp. Pharmak.*, ed. Erspamer. Berlin: Springer-Verlag.

Euler, U. S. von (1959). Autonomic neuroeffector transmission. In *Handbook of Physiology*, vol. I, Section I, Neurophysiology, p. 215. Washington, D.C. American Physiology Society.

Falck, B. (1962). Observations on the possibilities of the cellular localization of monoamines by a fluorescence method. *Acta physiol. scand.* **56,** Suppl. 197.

Glowinski, J. & Baldessarini, R. (1966). Brain catecholamines. *Pharmac. Rev.* (In the Press.)

Levi-Montalcini, R. (1964). Symposium on the nerve growth factor. *Ann. N.Y. Acad. Sci.* **118**, 147–232.

Weiner, N. (1964). The catecholamines: biosynthesis, storage and release, metabolism and metabolic effects. In *The Hormones*, vol. IV, New York: Academic Press.

Whittaker, V. P. (1965). The application of subcellular fractionation techniques to the study of brain function. *Prog. Biophys. Molec. Biol.* **15**, 41–96.

Zaimis, E. (1964). Pharmacology of the autonomic nervous system. *A. Rev. Pharmac.* **4**, 365–400.

2

CATECHOLAMINE METHODOLOGY

The recent upsurge of interest and research activity in fields related to the catecholamines can largely be attributed to the introduction of new techniques for the chemical isolation and estimation of noradrenaline and adrenaline and to the availability of radioactively labelled adrenaline and noradrenaline of high specific activity. In this chapter, the techniques most generally used in catecholamine research will be briefly surveyed. It is not intended that each method should be described in detail; for such information, original articles or more comprehensive text-books should be consulted. On the other hand, it seems worth while to devote this chapter to a critical survey of methods since the extensive array of techniques available can cause confusion.

A. METHODS FOR THE EXTRACTION AND ISOLATION OF CATECHOLAMINES FROM ANIMAL TISSUES AND FLUIDS

1. Extraction from tissues

In view of the stability of catecholamines in tissues kept at room temperature (Hökfelt, 1951) it is not generally considered necessary to take special precautions for the rapid freezing of tissues before an extract is prepared. Animal tissues or urine can be stored at low temperatures (−15 °C) for several weeks without loss of catecholamines. Following the introduction of perchloric acid and a report of its advantages over other agents (Bertler, Carlsson & Rosengren, 1958), the use of 0·4 N perchloric acid as an extracting medium has been widely adopted.

2. Isolation of catecholamines by alumina adsorption

The most widely used method for the isolation of noradrenaline and adrenaline from biological sources has been by adsorption of the catecholamines on aluminium oxide. The important discovery that catecholamines are strongly adsorbed on aluminium hydroxide at alkaline pH (8–8·5) but are not adsorbed at acid pH was made by Shaw (1938). This finding was subsequently developed as a method for the isolation of catecholamines from blood, urine or tissue extracts by many workers. Current methods are based on modifications of the technique described by Weil-Malherbe & Bone (1952). Adsorption is generally carried out at a pH of 8·4–8·6 using aluminium oxide (alumina). Adsorption may be effected by shaking or stirring the sample with alumina, or by pouring the sample through a bed of alumina in a small glass column. Because of the rapid spontaneous oxidation of catecholamines in alkaline media it is necessary to add a combination of a metal chelating agent (EDTA) and an antioxidant (ascorbic acid or sodium metabisulphite) to the sample before adjusting to pH 8·4–8·5. After adsorption of the sample, the alumina is washed with water or with a buffer solution and the catecholamines are then eluted in a small volume of dilute acid (0·2M acetic or hydrochloric acids are commonly used). Recent modifications of the alumina method which have gained widespread acceptance are described by Whitby, Axelrod & Weil-Malherbe (1961) and by Anton & Sayre (1962). The alumina adsorption method is specific for compounds with the catechol grouping, so that among the catecholamines and their metabolites only DOPA, dopamine, noradrenaline and adrenaline and their catechol metabolites—3,4-dihydroxy mandelic acid and 3,4-dihydroxyphenylacetic acid—and the corresponding glycols are isolated. Since the latter metabolites are generally of only minor quantitative importance, the method is adequately specific for most purposes. However, recoveries of catecholamines isolated by this technique are often not very high (60–80 %) and may vary considerably with different batches of alumina. Unless special precautions are taken in pretreatment of the alumina difficulties may also be encountered in the subsequent fluorimetric assay of noradrenaline or other catecholamines in the purified eluates. High blank readings and substantial quenching of internal standards of noradrenaline in the fluorimetric assay of noradrenaline in alumina eluates are common occurrences (Sharman, Vanov & Vogt, 1962; Iversen, 1963). For the isolation of radioactively labelled catecholamines, where no fluorimetric assay is involved, the alumina adsorption technique is the method of choice.

3. Isolation of catecholamines on ion-exchange resins

Bergström & Hansson (1951) reported the use of the weak acid resin Amberlite IRC-50 for the isolation of noradrenaline and adrenaline, and Kirshner, Goodall & Rosen (1959) also used this resin for the isolation and separation of catecholamines and their basic metabolites from urine. An ion-exchange resin procedure for the extraction of catecholamines from tissue extracts was described by Bertler *et al.* (1958); these and most subsequent workers used a strong acid resin of the sulphonated polystyrene type, with low cross-linkage (e.g. Dowex 50—X4). More recently strong acid resins of this type have been used for the chromatographic separation of noradrenaline and adrenaline from their *O*-methylated metabolites (Häggendal, 1962*a,b*). The use of strong acid resins is a highly specific method for the isolation of catecholamines from tissue extracts or other sources. At neutral pH only the catecholamines and their *O*-methylated metabolites are retained on the resin, all other related acidic or neutral compounds are removed in the column effluent and washings. By using a suitably long resin column, a resin of small particle size and by adjusting the strength and volumes of acid used for elution a virtually complete separation of noradrenaline and adrenaline from their *O*-methylated metabolites and dopamine can be achieved (Häggendal, 1962*b*; Iversen, 1963). On the other hand, if such a separation is not essential, a simpler procedure using a shorter resin column can be used (Bertler *et al.* 1958; Iversen, 1963). The recoveries of noradrenaline, adrenaline and dopamine from strong acid resins are generally high (85–95 %) and easily reproducible. The purified eluates obtained from resin columns are also particularly suitable for the subsequent fluorimetric analysis of the catecholamines since low blanks and little or no quenching are generally obtained.

B. CHROMATOGRAPHIC SEPARATION METHODS

Paper chromatographic methods for the separation of the catecholamines and their metabolites have been described by many workers (James, 1948; Crawford, 1951; James & Kilbey, 1950). Paper chromatography of concentrated extracts of tissues or plasma has been used as a preparative method for the isolation and separation of noradrenaline and adrenaline prior to their bioassay (Vogt, 1952). A useful modification has been the technique of acetylating the catecholamines prior to chromatography (Goldstein, Friedhoff & Simmons, 1959). Unlike the catecholamines themselves, the acetylated derivatives are relatively stable and can thus be chromatographed without danger of breakdown on the paper. Excellent separations of the various acetylated products can also be achieved with this method. The use of cellulose phosphate paper as a convenient method for the separation of catecholamines from their acidic or neutral metabolites has been described by Roberts (1962).

Catecholamines and their metabolites can also be separated by partition chromatography on starch or cellulose columns (Hamberg & Euler, 1950; Dengler, Michaelson, Spiegel & Titus, 1962).

C. THE FLUORIMETRIC ASSAY OF CATECHOLAMINES

1. Introduction

It is now generally recognized that fluorimetric methods are the only chemical techniques of sufficient sensitivity for the estimation of the very small amounts of adrenaline or noradrenaline normally encountered in animal tissues and body fluids. The development of reliable fluorimetric assay methods is a recent phenomenon and much of the existing knowledge about catecholamines was obtained by the use of bioassay procedures. Bioassay procedures remain the most sensitive methods for the estimation of noradrenaline and adrenaline and their use is still indicated when dealing with extremely small amounts of these substances, particularly since the shortcomings of the fluorimetric methods are most apparent under these conditions.

There are at present two chemical procedures for the estimation of catecholamines after their conversion into fluorescent derivatives. The trihydroxyindole method involves oxidation and cyclization and is the most commonly used procedure. An alternative method involves oxidation and condensation with ethylene diamine. These methods are described in detail by Crout (1961), Udenfriend (1962) and Weil-Malherbe (1961) and will be dealt with only briefly here.

2. The trihydroxyindole method

(*a*) *Introduction*

The fluorescence of adrenaline in alkaline solutions was first reported by Loewi in 1918, and later by Paget (1930). In 1934 Gaddum & Schild described a fluorimetric estimation of adrenaline based on its fluorescence under alkaline conditions, and this was extended for the assay of adrenaline in blood by Hueber (1940). These methods were extremely unreliable as the fluorescence of adrenaline in alkaline solution fades very rapidly, and is not of sufficient intensity to allow the detection of small amounts of adrenaline. Noradrenaline gave only 1–2 % of the fluoresence of adrenaline (Gaddum & Schild, 1934).

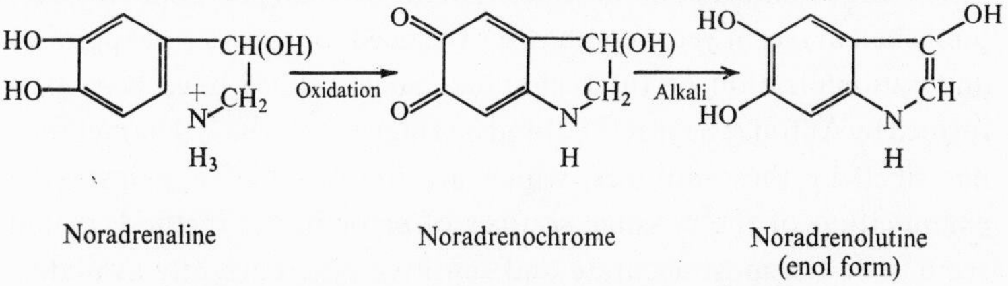

Noradrenaline —Oxidation→ Noradrenochrome —Alkali→ Noradrenolutine (enol form)

Fig. 2.1. Chemistry of the trihydroxyindole reaction.

It was not until chemical studies (Ehrlén, 1948; Lund, 1949; Harley-Mason, 1950) had elucidated the mechanism of the oxidation and rearrangement of catecholamines in alkaline solutions to form the fluorescent trihydroxyindole derivatives that satisfactory applications of these reactions became available. The first reliable method based on the formation of the trihydroxyindole derivatives was described by Lund in 1949. The principle of this and all the subsequent modifications of this method is as follows: adrenaline and noradrenaline are oxidized at slightly acid pH to form their quinone derivatives, adrenochrome and noradrenochrome; in alkaline solution these undergo intramolecular rearrangement to form the cyclized trihydroxyindoles, adrenolutine and noradrenolutine. These derivatives are unstable in alkaline solution in the presence of oxidizing agents, but can be stabilized by the addition of a reducing agent, ascorbic acid, when solutions are made alkaline. The reactions thought to occur are shown for noradrenaline in Fig. 2.1. It should be stated, however, that the chemistry of the trihydroxyindole procedure is not fully understood. Whilst there is little doubt that the reactions described do occur, the role of ascorbic acid in the later stages remains obscure. It does not seem to be acting simply as a reducing agent since other reducing agents, such as sulphite or thiosulphate, are unable to replace ascorbic acid in this procedure. Recent evidence (Harrison, 1963) has shown that ascorbic acid itself reacts with an intermediate oxidation product of noradrenaline to form a fluorescent compound, and the possible involvement of ascorbic acid in such a reaction in the trihydroxyindole procedure cannot be ruled out.

(*b*) *Modifications*

The large number of modifications of the original procedure of Lund published in recent years emphasizes the need for a careful reappraisal and rationalization of the technique. Such studies have been performed by Anton & Sayre (1962) and by Häggendal (1963). The methods described by these authors, which are the result of a painstaking examination of the possible sources of error in the method, would seem to be the most accurate and sensitive ones currently available.

Some of the more important modifications of the trihydroxyindole procedure will be briefly described. Lund (1949) recommended the use of manganese dioxide as the oxidizing agent, but most subsequent procedures have adopted iodine (Shore & Olin, 1958; Crout, 1961) or ferricyanide (Price & Price, 1957; Bertler *et al.* 1958; Vendsalu, 1960; Euler & Floding, 1955). Shore & Olin (1958) claimed that oxidation with iodine produced a product with twice the fluorescence of that obtained after ferricyanide oxidation, but this finding has not been confirmed by other workers. In view of the relatively unspecific nature of the iodine oxidation, the light-sensitivity and the long interval (30–60 min) required for its completion, ferricyanide would seem to be the more satisfactory oxidizing agent. During iodine treatment it seems clear that other chemical reactions occur apart from oxidation, forming for instance the iodochrome derivatives of adrenaline and noradrenaline. Iodine also has the disadvantage that it forms a fluorescent product from dopamine which may well be present in many tissue extracts or urine samples. On the other hand, oxidation with ferricyanide is rapid (2–3 min) and relatively specific, thus dopamine is not appreciably oxidized and does not interfere with the assay of adrenaline and noradrenaline, even if present in very large amounts (Wegmann, 1963).

There have been numerous minor modifications concerning the strength of the sodium hydroxide solution or the relative amount of alkali which should be added (Euler & Lishajko, 1961; Anton & Sayre, 1962). Almost all methods, however, retain the addition of ascorbic acid to the alkaline solution in order to stabilize the lutines. Merrills (1962) reported an adaptation of the trihydroxyindole method for use with an automatic analyser; this procedure introduced the use of thioglycollic acid in place of ascorbic acid as a stabilizing agent. This modification had the advantage that it was claimed to make the method specific for noradrenaline; i.e. thioglycollic acid stabilized the fluorescent product of noradrenaline but not that of adrenaline. Palmer (1963) reported that β-thiopropionic acid could also be used as an alternative to ascorbic acid and in the method of Häggendal (1963) a mixture of BAL and sodium sulphite is used in place of ascorbic acid.

(*c*) *Differential assay of adrenaline and noradrenaline*

The original method of Lund (1949) was designed to allow a differential assay of adrenaline and noradrenaline in a mixture of the two compounds. Whilst both adrenaline and noradrenaline are completely oxidized by manganese dioxide within a few minutes at pH 6·0, at pH 3·0 the oxidation of adrenaline is much faster than that of noradrenaline. This allows a differential assay to be carried out by oxidizing samples at both pH values for carefully controlled intervals before the addition of alkali/ascorbic to each. The noradrenaline and adrenaline content of the original sample can then be calculated from the relative amounts of fluorescence produced at each pH. This principle has been adopted in many subsequent differential methods, using iodine or ferricyanide oxidation at two pH values. The pH values generally used are 3·5 and 6·5. These procedures give ratios of adrenaline to noradrenaline fluorescence at the lower pH as high as 50:1 (Table 2.1) and are thus capable of giving a very sensitive discrimination of the two amines. They suffer from the disadvantages that the pH and duration of each oxidation has to be very carefully controlled, and the readings have to be taken from two different samples, thus entailing the preparation of twice as many experimental samples and blanks for the assay procedure.

Alternative differential assay procedures depend on the differential fluorescence of noradrenaline and adrenaline at different activating and exciting wavelengths in a spectrophotofluorimeter, or using differ-

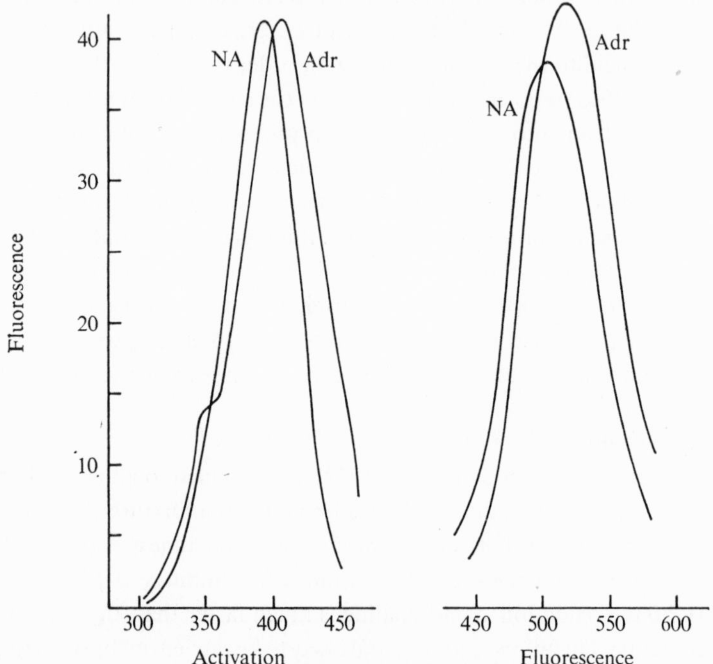

Fig. 2.2. Fluorescence spectra of the lutines formed from noradrenaline (NA) and adrenaline (Adr) in the trihydroxyindole reaction. Fluorescence in arbitrary units; wavelengths in mμ (uncorrected instrument values).

ent filter sets in a filter fluorimeter. These methods, suggested by Price & Price (1957) and Cohen & Goldenberg (1957), depend on differences in the fluorescence spectra of adrenolutine and noradrenolutine. The spectra are shown in Fig. 2.2. Differential assays by this method involve reading the same sample of a noradrenaline/adrenaline mixture at two different wavelength combinations in a spectrophotofluorimeter; for example: Bertler *et al.* (1958) for noradrenaline, exciting = 410, fluorescence = 540; for adrenaline, exciting = 455, fluorescence =

Table 2.1. *Comparison of differential methods for the assay of adrenaline and noradrenaline*

(After Crout, 1961.)

Principle	Oxidizing agent	Ratio A/NA fluorescence pH 6·5	pH 3·5	Reference
Differential pH method	Iodine	0·5/1	50/1	Crout, 1961
	Ferricyanide	1·6/1	25/1	Euler & Floding, 1955
	MnO_2	1·4/1	20/1	Lund, 1949
		Filter (wavelength) combination A	B	
Differential filter (or wavelength) method	Ferricyanide	1/1	3/1	Euler & Lishajko, 1961
	Ferricyanide	1·2/1	5·7/1	Bertler *et al.* 1958
	MnO_2	1·1/1	5·8/1	Cohen & Goldenberg, 1957

540 (wavelengths in mμ). In filter instruments various combinations of primary and secondary filter sets have been used; most use exciting light from the 405 mμ band for adrenaline and secondary filters for the fluorescent light in the range 500–540 mμ. In all cases the calculation of adrenaline and noradrenaline in a mixture involves solving a simultaneous equation in which constants describing the properties of pure noradrenaline and adrenaline under the two recording conditions are incorporated. Under the best possible conditions, i.e. using a spectrophotofluorimeter, a maximum ratio of adrenaline/noradrenaline of 6:1 can be achieved, and most filter instrument methods give a much smaller differentiation than this. Table 2.1 summarizes the results obtained in differential assays by both types of procedure. The differential pH method is clearly the more sensitive of the two procedures, but its other disadvantages have generally led to the use of the differential wavelength procedures, which have the advantage that the differential assay can be carried out on a single sample and is not critically dependent on the maintenance of carefully controlled conditions of oxidation.

(d) The preparation of blanks

The preparation of reagent blanks is complicated by the fact that solutions of ascorbic acid in strong alkali develop a weak fluorescence on standing, at the same time the solutions assume a faint pink colour. The chemical reactions involved in this phenomenon are not clearly understood. Euler & Lishajko (1961) reported that this development of fluorescence in reagent blanks could be suppressed by the addition of 2 % ethylene diamine to the alkaline/ascorbic mixture and this useful modification has been widely accepted.

In the fluorimetric assay of catecholamines in biological extracts it is clearly also important to include sample blanks to correct for interfering compounds which may be present in the experimental samples. For this purpose a 'faded' blank is generally used: the extract is oxidized in the normal manner, alkali is then added *without* ascorbic acid; the noradrenaline and adrenaline lutines fade rapidly under these conditions and after an interval of from 10–20 min ascorbic is added and the blanks read; any fluorescence is then assumed to derive from substances other than adrenaline or noradrenaline.

The presence of quenching substances in the extracts should be assessed by the inclusion of internal standards. Samples are prepared in duplicate; to one sample a known amount of adrenaline or noradrenaline is added and the recovery of this is then compared with the recovery of noradrenaline and adrenaline from pure aqueous solutions.

(e) Specificity

Chemical studies on the specificity of the trihydroxyindole reaction suggested that it requires the presence of a catechol nucleus, a β-OH substituent and an alkyl amine substituent on the α-carbon (Bu'lock & Harley Mason, 1951; Lund, 1949).

Of the naturally occurring compounds of this type only noradrenaline and adrenaline are of any importance; however, substances lacking a β-hydroxyl may undergo similar reactions to yield dihydroxyindole derivatives with similar fluorescence properties to noradrenolutine and adrenolutine, thus both DOPA and dopamine interfere to some extent with the assay. If ferricyanide is used as the oxidant, the interference by dopamine is negligible, amounting to less than 2 % of the fluorescence derived from noradrenaline (Anton & Sayre, 1962). DOPA on the other hand gives some 12 % of the noradrenaline fluorescence, and the trihydroxyindole procedure can be used without modification for the assay of this compound (Wegmann, Kako & Chrysohuo, 1963; Bertler *et al.* 1958). In this procedure DOPA is separated from noradrenaline, adrenaline and dopamine by passage through a column of Amberlite CG-50 followed by adsorption of DOPA from the column effluent on alumina. Modifications of the trihydroxyindole procedure have been described which allow the assay of other related compounds, such as dopamine (Carlsson & Waldeck, 1958; Drujan, Sourkes, Layne & Murphy, 1959; McGeer & McGeer, 1962) or normetanephrine and metanephrine (Bertler, Carlsson & Rosengren, 1959; Häggendal, 1962*b*).

Certain pharmacologically active substances closely related in structure to the catecholamines will also yield fluorescent products in the trihydroxyindole procedure; this may be of importance when these drugs are used, either as an assay method for the drug itself, or as an interfering factor in the determination of adrenaline and noradrenaline. The following drugs were found by Anton & Sayre (1962) to yield fluorescent products, *N*-ethyl-noradrenaline, isoprenaline (*c.* 50 % noradrenaline fluorescence), α-methyl DOPA (17·2 %), α-methyl-noradrenaline (*c.* 10 %), *N*-methyl-dopamine (1 %).

3. The ethylene diamine condensation method

The second procedure which has been used for the fluorimetric assay of catecholamines is condensation with ethylene diamine. The intermediate quinone oxidation products of adrenaline and noradrenaline react with ethylene diamine to form highly fluorescent products. The chemistry of these condensations is complex and still only partially understood. The application of this reaction for the estimation of catecholamines was first described by Weil-Malherbe & Bone (1952). Although the fluorescent products of noradrenaline and adrenaline formed by this method have a more intense fluorescence than those formed by the trihydroxyindole method, the ethylene diamine condensation is far less specific than the tryhydroxyindole reaction, and thus less useful. Ethylene diamine condenses with many catechol compounds to give fluorescent products and this lack of specificity undoubtedly accounts for the high plasma levels of adrenaline and noradrenaline reported by workers using this method. On the other hand, by suitably purifying the catecholamine extracts it is possible to use the ethylene diamine reaction as a relatively specific and highly sensitive procedure. After several modifications to the original method Weil-Malherbe (1961) has greatly increased the specificity of the procedure. The revised method involves the extraction of noradrenaline and adrenaline from the sample by passage through an ion-exchange resin column; the catecholamines are then eluted and further purified by passage through a column of alumina to remove non-catechol amines. The ethylene diamine reaction also has other possible uses, for example for the fluorimetric assay of catechol-acids (Rosengren, 1960) or dopamine (Carlsson & Waldeck, 1958). A highly specific and sensitive assay for dopamine has recently been described in which the catecholamines in tissue extracts are acetylated and extracted into dichloromethane; the acetylated derivatives are separated by paper chromatography, eluted from the paper, hydrolysed and condensed with ethylene diamine to form the fluorescent condensation products (Laverty & Sharman, 1965).

D. HISTOCHEMICAL TECHNIQUES

The demonstration of catecholamines in cells of the adrenal medulla and in other chromaffin tissues by the chromaffin reaction of such cells has been used for many years. However, it is only recently that histochemical techniques have become available which are sufficiently sensitive to demonstrate the presence of the very small amounts of catecholamines normally present in adrenergic neurones. Eränkö (1955) was the first to describe the use of formaldehyde condensation reactions to visualize catecholamines in the cells of the adrenal medulla. Hillarp & Falck and their co-workers made the important discovery that catecholamines in a dry protein matrix would condense with formaldehyde vapours to yield highly fluorescent derivatives, although this reaction did not occur in the absence of protein or in the presence of excessive amounts of water (Falck, Hillarp, Thieme & Thorpe, 1962). This principle was found to be applicable to catechol-

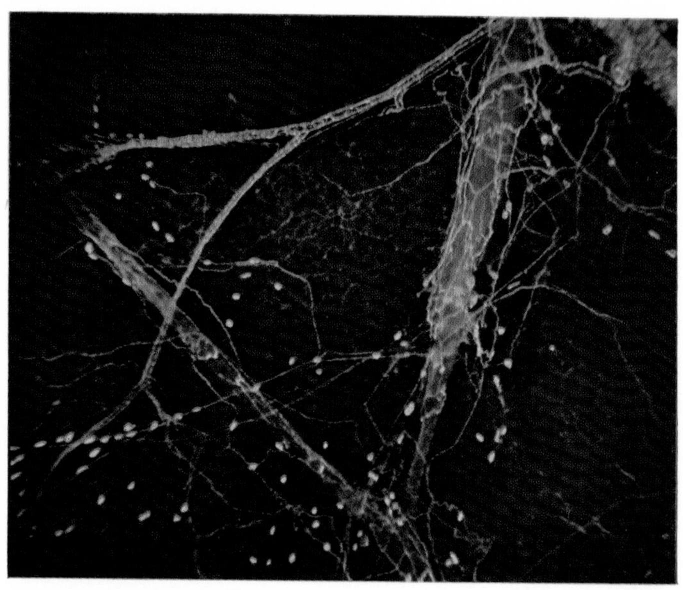

1 Fluorescence micrograph of monoamines after treatment with formaldehyde according to the technique of Falck & Owman (1965). Stretch preparation of rat mesenterium. A yellow fluorescence, due to 5-hydroxytryptamine, is seen in numerous mast cells, while a green noradrenaline fluorophore is present in adrenergic nerve terminals forming adrenergic ground plexa around arterioles and small arteries (the vascular walls are outlined by a greenish unspecific background fluorescence), or running isolated between the vessels. Magnification ×150. From C. Owman, unpublished.

amines in freeze-dried tissue sections, allowing the localization of the catecholamines in such sections by fluorescence microscopy after formaldehyde treatment (Falck, 1962). The fluorescence histochemical technique described by Falck represents a major advance in catecholamine methodology and has gained widespread and rapid application. The method is described in detail by Dahlström & Fuxe (1964) and by Falck & Owman (1965). When catecholamines in freeze-dried tissue sections are exposed to formaldehyde vapours at 80 °C, the highly fluorescent 6,7-dihydroxy-3,4-dihydroisoquinoline derivatives are formed (Corrodi & Hillarp, 1963, 1964). By illuminating the tissue sections with ultraviolet light and observing the green or green-yellow fluorescence of these derivatives in a fluorescence microscope it is possible to visualize the catecholamines (Plate 1). Indoleamines, such as serotonin, also react to form fluorescent derivatives, but these can be distinguished from those of the catecholamines by virtue of their different fluorescence characteristics. Adrenaline and other secondary catecholamines can also be distinguished from primary catecholamines such as noradrenaline, dopamine and DOPA because the reaction of adrenaline with formaldehyde requires much longer for completion. It has so far not been possible to distinguish between the primary catecholamines dopamine and noradrenaline, since the fluorescent products from these two amines have almost identical fluorescence spectra. The 4,6,7-trihydroxy-3,4-dihydroisoquinoline formed from noradrenaline, however, has a labile hydroxy group at the 4-position which can be easily split off. The fully aromatic 6,7-dihydroxyisoquinoline thus formed has different fluorescence characteristics, so that it may be possible to distinguish between dopamine and noradrenaline in this way (Corrodi & Jonsson, 1965). The method now generally used is highly sensitive and specific for the localization of catechol- and indole-amines; the following criteria have been recommended to distinguish between specific and non-specific fluorescence in formaldehyde-treated samples (Dahlström & Fuxe, 1964; Norberg & Hamberger, 1964; Corrodi, Hillarp & Jonsson, 1964):

(1) A marked reduction or complete disappearance of specific fluorescence when sections are exposed to water.

(2) An almost complete disappearance of specific fluorescence on reduction with sodium borohydride.

(3) Absence of specific fluorescence in samples from animals pretreated with the amine-depleting drug reserpine.

(4) Increased specific catecholamine fluorescence after treatment with a monoamine oxidase inhibitor and DOPA.

E. THE USE OF RADIOACTIVELY LABELLED MATERIALS

The availability of tritiated noradrenaline and adrenaline of high specific activity (> 3 c/m.mole) and the advent of liquid scintillation counting systems for the efficient detection of tritium have been major factors in the recent advances in this field. Because of the high specific

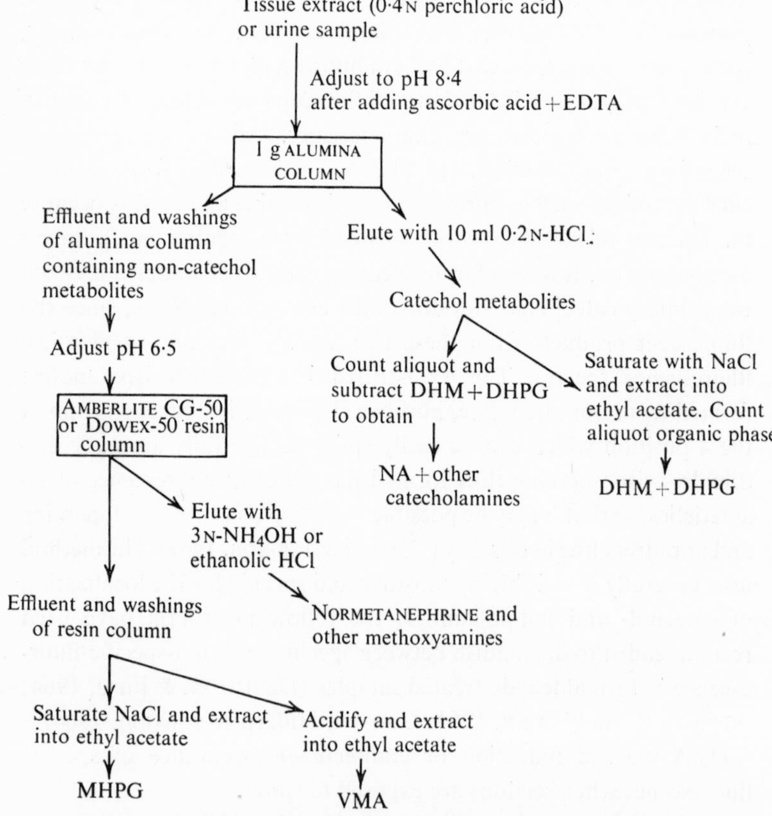

Fig. 2.3. Outline of methods for separating radioactive catecholamine metabolites (after Kopin, Axelrod & Gordon, 1961).

activity of the labelled catecholamines it is possible to detect extremely small amounts of these substances in tissue extracts or body fluids—as little as 10 pg of H^3—noradrenaline can be accurately measured in a single sample. The methods used for the isolation of radioactively labelled catecholamines from biological sources are the same as those described above. It should be emphasized that the use of isotopically labelled materials makes it essential that great care should be taken in the isolation of labelled catecholamines or their metabolites in a pure form prior to assay. Methods for the isolation of the catecholamines and their various individual metabolites have been described by Kopin, Axelrod & Gordon (1961), and are illustrated in outline in Fig. 2.3.

The recent introduction of isotopically labelled catecholamine precursors of high specific activity has opened new possibilities for the formation of labelled catecholamines in tissue stores by administering these substrates for biosynthesis. The following compounds are currently available with a specific activity of more than 100 mc/m-mole: DL-noradrenaline, DL-adrenaline, DL-isoprenaline, dopamine, L-DOPA, L-tyrosine, tyramine.

REFERENCES

Anton, A. H. & Sayre, D. F. (1962). A study of the factors affecting the aluminium oxide trihydroxyindole procedure for the analysis of catecholamines. *J. Pharmac. exp. Ther.* **138,** 360–375.

Bergström, S. & Hansson, G. (1951). The use of Amberlite IRC-50 for the purification of adrenaline and histamine. *Acta physiol. scand.* **22,** 87–92.

Bertler, A., Carlsson, A. & Rosengren, E. (1958). A method for the fluorimetric determination of adrenaline and noradrenaline in tissues. *Acta physiol. scand.* **44,** 273–292.

Bertler, A., Carlsson, A. & Rosengren, E. (1959). Fluorimetric method for differential estimation of the 3-0-methylated derivatives of adrenaline and noradrenaline (metanephrine and normetanephrine). *Clinica chim. Acta* **4,** 456–457.

Bu'lock, J. & Harley-Mason, J. (1951). The chemistry of adrenochrome. II. Some analogues and derivatives. *J. chem. Soc.* 712–716.

Carlsson, A. & Waldeck, B. (1958). Fluorimetric method for the determination of dopamine (3-hydroxytyramine). *Acta physiol. scand.* **44,** 293–298.

Cohen, G. & Goldenburg, M. (1957). The simultaneous fluorimetric determination of adrenaline and noradrenaline in plasma. I. The fluorescence characteristics of adrenolutine and noradrenolutine and their simultaneous determination in mixtures. *J. Neurochem,* **2,** 58–70.

Corrodi, H. & Hillarp, N.-Å. (1963). Fluoreszenzmethoden zur histochemischen Sichtbarmachung von Monoaminen. 1. Identifizierung der fluoreszierenden Produkte aus Modellversuchen mit 6,7-dimethoxy-isochinolinderivaten und Formaldehyd. *Helv. chim. Acta* **46**, 2425–2430.

Corrodi, H. & Hillarp, N.-Å. (1964). Fluoreszenzmethoden zur histochemischen Sichtbarmachung von Monoaminen. 2. Identifizierung des fluoreszierenden Produktes aus Dopamin und Formaldehyd. *Helv. chim. Acta* **47**, 911–918.

Corrodi, H., Hillarp, N.-Å. & Jonsson, G. (1964). Fluorescence methods for the histochemical demonstration of monoamines. Sodium borohydride reduction of the fluorescent compounds as a specificity test. *J. Histochem. Cytochem.* **12,** 582–586.

Corrodi, H. & Jonsson, G. (1965). Fluorescence methods for the histochemical demonstration of monoamines. 4. Histochemical differentiation between dopamine and noradrenaline in models. *J. Histochem. Cytochem.* **13,** 484–489.

Crawford, T. B. (1951). Derivatives of adrenaline and noradrenaline in an extract of an adrenal medullary tumour. *Biochem. J.* **48**, 203–208.

Crout, J. R. (1961). In *Standard Methods in Clinical Chemistry* (ed. Seligson), vol. III, p. 62. New York: Academic Press.

Dahlström, A. & Fuxe, K. (1964). Evidence for the existence of monoamine-containing neurones in the central nervous system. *Acta physiol scand.* **62,** Suppl. 232.

Dengler, H. J., Michaelson, I. A., Spiegel, H. E. & Titus, E. O. (1962). The uptake of labelled norepinephrine by isolated brain and other tissues of the cat. *Int. J. Neuropharmacol.* **1,** 23–38.

Drujan, B. D., Sourkes, T. A., Layne, D. S. & Murphy, G. F. (1959). The differential determination of catecholamines in urine. *Can. J. Biochem. Physiol.* **37,** 1153–1159.

Ehrlén, I. (1948). Fluorimetric determination of adrenaline. II. *Farmaceutisk Revy* **47,** 242–250.

Eränkö, O. (1955). The histochemical demonstration of noradrenaline in the adrenal medulla of rats and mice. *J. Histochem. Cytochem.* **4,** 11.

Euler, U. S. von & Floding, I. (1955). Fluorimetric estimation of noradrenaline and adrenaline in urine. *Acta physiol. scand.* **33,** Suppl. 118, p. 57.

Euler, U. S. von & Lishajko, F. (1961). Improved technique for the fluorimetric estimation of catecholamines. *Acta physiol. scand.* **51,** 348–355.

Falck, B. (1962). Observations on the possibility of the cellular localization of monoamines by a fluorescence method. *Acta physiol. scand.* **56,** Suppl. 197.

Falck, B., Hillarp, N.-Å., Thieme, G. & Thorpe, A. (1962). Fluorescence of catecholamines and related compounds condensed with formaldehyde. *J. Histochem. Cytochem.* **10**, 348–354.

Falck, B. & Owman, C. (1965). A detailed methodological description of the fluorescence method for the cellular demonstration of biogenic amines. *Acta Univ. lund.* Sectio II, no. 7.

Gaddum, J. H. & Schild, H. (1934). A sensitive physical test for adrenaline. *J. Physiol., Lond.* **80**, 9–10*P*.

Goldstein, M., Friedhoff, A. J. & Simmons, C. (1959). A method for the separation and estimation of catechol amines in urine. *Experientia* **15,** 80–81.

Häggendal, J. (1962*a*). On the use of strong exchange resins for determinations of small amounts of catechol amines. *Scand. J. clin. Lab. Invest.* **14**, 537–544.

Häggendal, J. (1962*b*). Fluorimetric determination of 3-0-methylated derivatives of adrenaline and noradrenaline in tissues and body fluids. *Acta physiol. scand.* **56**, 258–266.

Häggendal, J. (1963). An improved method for fluorimetric determination of small amounts of adrenaline and noradrenaline in plasma and tissues. *Acta physiol. scand.* **59**, 242–254.

Hamberg, U. & Euler, U. S. von. (1950). Partition chromatography of adrenaline and noradrenaline. *Acta chem. scand.* **4**, 1185–1191.

Harley-Mason, J. (1950). The chemistry of adrenochrome and its derivatives. *J. Chem. Soc.* 1276–1282.

Harrison, D. (1963). Ascorbic acid induced fluorescence of a noradrenaline oxidation product. *Biochim. biophys. Acta* **78,** 705–710.

Hökfelt, B. (1951). Noradrenaline and adrenaline in mammalian tissues. Distribution under normal and pathological conditions with special reference to the endocrine system. *Acta physiol. scand.*, **25**, Suppl. 92.

Hueber, H. F. (1940). Über eine einfache Methode, Adrenaline im strömenden Blut Nachzuweissen. *Klin. Wschr.* **19**, 664–665.

Iversen, L. L. (1963). The uptake of noradrenaline by the isolated perfused rat heart. *Br. J. Pharmac. Chemother.* **21**, 523–537.

James, W. O. (1948). Demonstration and separation of noradrenaline, adrenaline and methyladrenaline. *Nature, Lond.* **161**, 851–852.

James, W. O. & Kilbey, N. (1950). Separation of noradrenaline and adrenaline. *Nature, Lond.* **166,** 67–68.

Kirshner, N., Goodall, McC. & Rosen, L. (1959). The effect of iproniazid on the metabolism of DL-epinephrine-2-C^{14} in the human. *J. Pharmac. exp. Ther.* **127,** 1–7.

Kopin, I. J., Axelrod, J. & Gordon, E. (1961). The metabolic fate of H^{3}-epinephrine and C^{14}-metanephrine in the rat. *J. biol. Chem.* **236**, 2109–2113.

Laverty, R. & Sharman, D. F. (1965). The estimation of small quantities of 3,4-dihydroxyphenylethylamine in tissues. *Br. J. Pharmac.* **24,** 538–548.

Loewi, O. (1918). Über die Natur der Giftwirkung des Suprarenins. *Biochem. Z.* **85**, 295–306.

Lund, A. (1949). Fluorimetric determination of adrenaline in blood, I, II, III. *Acta pharmac. tox.* **5,** 75–94, 121–128, 231, 247.

McGeer, E. G. & McGeer, P. L. (1962). Catecholamine content of spinal cord. *Can. J. Biochem. Physiol.* **40,** 1141–1151.

Merrills, R. J. (1962). An autoanalytical method for the estimation of adrenaline and noradrenaline. *Nature, Lond.* **193**, 988.

Norberg, K. A. & Hamberger, B. (1964). The sympathetic adrenergic neurone. *Acta physiol. scand.* **63**, Suppl. 238.

Paget, M. (1930). Nouvelle réaction colorée de l'adrénaline et de l'adrénalone. *Bull. Sci. Pharmac.* **37**, 537–538.

Palmer, J. F. (1963). The use of β-thiopropionic acid for stabilising the fluorescence of adrenaline and noradrenaline. *J. Pharm. Pharmac.* **15**, 777–778.

Price, H. L. & Price, M. L. (1957). The chemical estimation of epinephrine and norepinephrine in human and canine plasma. *J. Lab. clin. Med.* **50**, 769–777.

Roberts, M. (1962). A note on the use of cellulose phosphate cation-exchange paper for the separation of catecholamines, and some other biogenic amines. *J. Pharm. Pharmac.* **14**, 746–749.

Rosengren, E. (1960). On the role of monoamine oxidase for the inactivation of dopamine in brain. *Acta physiol. scand.* **49**, 370–375.

Sharman, D. F., Vanov, S. & Vogt, M. (1962). Noradrenaline content in the heart and spleen of the mouse under normal conditions and after administration of some drugs. *Br. J. Pharmac.* **19**, 527–533.

Shaw, F. H. (1938). The estimation of adrenaline. *Biochem. J.* **32**, 19–25.

Shore, P. A. & Olin, J. S. (1958) Identification and chemical assay of norepinephrine in brain and other tissues. *J. Pharmac. exp. Ther.* **122**, 295–300.

Udenfriend, S. (1962). *Fluorescence Assay in Biology and Medicine.* New York: Academic Press.

Vendsalu, A. (1960). Studies on adrenaline and noradrenaline in human plasma. *Acta physiol. scand.* **49**, Suppl. 173.

Vogt, M. (1952). The secretion of the denervated adrenal medulla of the cat. *Br. J. Pharmac. Chemother.* **7**, 325–330.

Wegmann, A. (1963). Determination of 3-hydroxytyramine and Dopa in various organs of the dog after Dopa-infusion. *Arch. exp. Path. Pharmak.* **246**, 184–190.

Wegmann, A., Kako, K. & Chrysohuo, A. (1963). Dopa uptake and catecholamine content in heart and spleen. *Am. J. Physiol.* **201**, 673–676.

Weil-Malherbe, H. (1961). The fluorimetric estimation of catecholamines. In *Methods in Medical Research* (ed. Quastel), vol. IX, pp. 130–146.

Weil-Malherbe, H. & Bone, A. D. (1952). The chemical estimation of adrenaline-like substances in blood. *Biochem. J.* **51**, 311–318.

Whitby, L. G., Axelrod, J. & Weil-Malherbe, H. (1961). The fate of H^3 norepinephrine in animals. *J. Pharmac. exp. Ther.* **132**, 193–201.

3

THE DISCOVERY OF THE SYMPATHETIC NEUROTRANSMITTER

The concept of chemical transmission at synapses was first proposed by Elliot in 1905, and later by Dixon & Hamill (1909). It was suggested that adrenaline might act at the junction of sympathetic nerves and the effector organs, and that adrenaline might be released at sympathetic nerve endings. The first experimental confirmation of this hypothesis was obtained by Loewi (1921) and in the same year by Cannon & Uridil (1921). Because of the similarities between the effects of sympathetic nerve stimulation and those of adrenaline it was assumed that the transmitter liberated at sympathetic nerve endings was adrenaline, and this view was to remain generally accepted until 1945. The great interest in the new hormone adrenaline in the early years of the century after its discovery and isolation from adrenal extracts (Oliver & Schäfer, 1895; Takamine, 1901) helped to establish the erroneous identity of adrenaline as the sympathetic transmitter substance. The subsequent findings that postganglionic stimulation of the nerves of the rabbit's ear released what appeared to be adrenaline (Gaddum & Kwiatkowski, 1939) and that certain organs and their sympathetic nerves seemed to contain adrenaline (Cannon & Lissák, 1939; Lissák, 1939; Loewi, 1936) all supported the original hypothesis. This view was so firmly accepted that it was not to be corrected for many years, and its acceptance led to much confusion in the field of adrenergic neurotransmission over that period.

Noradrenaline (Fig. 3.1) was synthesized as early as 1904 by Stolz, who also showed that it was less toxic and equally potent as a pressor substance when compared with adrenaline. The remarkable pressor activity and low toxicity of noradrenaline were also pointed out by Biberfeld (1906). Little serious attention, however, was paid at that time to the possibility that noradrenaline might play a physiological role. Nevertheless the discrepancies inherent in the concept of adrenaline as the sympathetic neurotransmitter were evident at a very early stage. Thus in 1910 Barger & Dale in their now classical paper on the actions of sympathomimetic amines noted that 'the actions of some of the other bases, particularly the amino- and amino-ethyl-bases of the catechol group [noradrenaline] corresponds more closely with that of sympathetic nerves than does that of adrenine [adrenaline]'. In particular Barger & Dale had noticed that the well-known reversal of the pressor response to adrenaline after blockade of the adrenergic receptors with ergotoxine was not found with noradrenaline or after

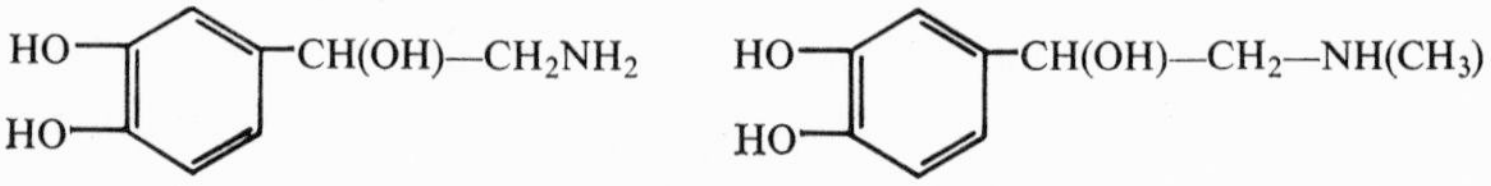

HO—C$_6$H$_3$(HO)—CH(OH)—CH_2NH_2 HO—C$_6$H$_3$(HO)—CH(OH)—CH_2—NH(CH_3)

Noradrenaline   Adrenaline

Fig. 3.1

stimulation of sympathetic nerves. Furthermore Cannon & Rosenblueth (1933) had noted certain differences between the action of the transmitter liberated by sympathetic nerve stimulation and that of adrenaline, and this led them to propose their theory of two sympathins. They suggested that a common mediator (M) was liberated from sympathetic nerve endings. This then combined with either substance E or substance I in the effector cell to form hypothetical compounds sympathin E or sympathin I which had excitatory and inhibitory activities respectively. Bacq (1934) suggested that noradrenaline might be equivalent to Cannon & Rosenblueth's sympathin E. Stehle & Ellsworth (1937) had again noted the closer similarity between the effects of sympathetic nerve stimulation and those of noradrenaline rather than adrenaline, and concluded that their results supported Bacq's view that noradrenaline was sympathin E.

By then, however, Bacq had modified his earlier view and had suggested (Bacq, 1935) that sympathin E might be a partially oxidized form of adrenaline rather than noradrenaline.

Another theory, which dispensed with the intermediate substances E and I in the effector tissue, was proposed by Greer, Pinkston, Baxter & Brannon (1938). After experiments which compared the results of injections of DL-noradrenaline, L-adrenaline and stimulation of the hepatic nerves in the cat before and after ergotoxine treatment, they suggested that sympathetic nerve endings might release either of two transmitters, which would then have direct actions on the effector cells. They suggested that the excitatory transmitter might be noradrenaline and the inhibitory transmitter adrenaline.

Another line of evidence which led to the final recognition of noradrenaline as a physiologically important substance was the discovery of the enzyme L-DOPA decarboxylase in mammalian tissues by Holtz, Heise & Lüdtke (1938). This enzyme was found to convert DOPA to dopamine, which differed from noradrenaline only in the absence of a β-hydroxyl group. That noradrenaline might be formed as an intermediate in mammalian tissues during the synthesis of adrenaline from DOPA was suggested by Blaschko in 1939.

The evidence that noradrenaline was the predominant catecholamine in peripheral tissues and adrenergic nerves came from the work of Euler (Euler, 1946*a*, *b*, *c*, 1948, 1950; Euler & Lishajko, 1957). A detailed comparison of extracts from splenic nerves and spleen (ox) and mammalian heart with adrenaline showed a number of differences and suggested that the catecholamine present in such extracts was predominantly noradrenaline. The evidence can be summarized as follows: (1) biological action on a wide variety of tests; (2) inhibition and potentiation of actions by dibenamine and cocaine; (3) colorimetric reactions with various oxidants; (4) (more recently) chromatographic behaviour, and fluorescence characteristics of the ethylene diamine condensation product. Euler concluded that noradrenaline was the predominant sympathetic neurotransmitter substance in these organs, though he retained the concept of Greer *et al.* (1938) that there might be two transmitter substances, and proposed that they be designated sympathin N (noradrenaline) and sympathin A (adrenaline).

The presence of noradrenaline in nerves and tissues was confirmed by Bacq & Fischer (1947); Holtz, Credner & Kroneberg (1947) demonstrated noradrenaline in human urine; Gaddum & Goodwin (1947) showed that noradrenaline was the transmitter in liver; Schmiterlöw (1948) demonstrated noradrenaline in extracts of the blood vessels of many species. Synthetic DL-noradrenaline was resolved into the optical isomers by Tullar (1948) and the naturally occurring material in splenic nerves was identified as the L-isomer of noradrenaline by Euler (1948).

After the demonstration that L-noradrenaline was present in organs innervated by the sympathetic nervous system came direct evidence that noradrenaline was released from such organs by postganglionic sympathetic nerve stimulation. This was first conclusively demonstrated by Peart (1949) after stimulation of the splenic nerves in the cat; and later by West (1950) for cat spleen; Mann & West (1950, 1951) for liver, spleen, uterus and gut; by Outschoorn & Vogt (1952) for dog heart and by Outschoorn (1952) for the perfused rabbit ear.

Despite the overwhelming evidence in favour of the identity of noradrenaline with the sympathetic neurotransmitter, the concept was not immediately accepted (e.g. critical review by Tainter & Luduena, 1950). Ahlquist (1948) postulated that adrenaline was the sole transmitter and that its action depended on the presence of two types of receptor termed α or β. The evidence in favour of the view that noradrenaline is the sole sympathetic transmitter has been reviewed by Euler (1951, 1956, 1958, 1959). The terms sympathin N and sympathin A are now not generally used; the current view is that noradrenaline is the only transmitter substance in adrenergic nerves in mammals, and that the traces of adrenaline found in many peripheral tissues are not connected with adrenergic neurotransmission (Euler, 1959).

Recently it has become possible for the first time to demonstrate the presence of noradrenaline in sympathetic nerve terminals directly, by use of the fluorescence histochemical method of Falck (1962). The use of this technique has confirmed that the noradrenaline present in peripheral mammalian tissues is localized almost exclusively in sympathetic nerve terminals (Falck & Torp, 1962). Falck, Häggendal & Owman (1963) have also demonstrated that the sympathetic nerve terminals of the frog heart contain high concentrations of adrenaline rather than noradrenaline, indicating that in some non-mammalian species adrenaline may be the sympathetic neurotransmitter. By a curious historical coincidence the original studies of Loewi (1921) with the isolated frog heart have thus been shown to be correct in the conclusion that adrenaline was the sympathetic neurotransmitter substance in that tissue.

REFERENCES

Ahlquist, R. P. (1948). A study of adrenotropic receptors. *Am. J. Physiol.* **153**, 586–600.

Bacq, Z. M. (1934). La pharmacologie du système nerveux autonome, et particulièrement du sympathique, d'après la théorie neurohumorale. *Ann. Physiol.* **10**, 467–528.

Bacq, Z. M. (1935). La transmission chemique des influx dans le système nerveux autonome. *Ergebn. Physiol.* **37**, 82–185.

Bacq, Z. M. & Fischer, P. (1947). Nature de la substance sympathicomimétique extraite des nerfs ou des tissus des mamifères. *Archs. int. Physiol.* **55**, 73–91.

Barger, G. & Dale, H. H. (1910). Chemical structure and sympathomimetic action of amines. *J. Physiol., Lond.* **41**, 19–59.

Biberfeld, J. (1906). Pharmakologische Eigenschaffen eines synthetisch dargestellten Suprarenins und einiger seiner Derivate. *Medsche. Klin.* 1177–1179.

Blaschko, H. (1939). The specific action of L-dopa decarboxylase. *J. Physiol., Lond.* **96**, 50–51*P*.

Cannon, W. B. & Lissák, K. (1939). Evidence for adrenaline in adrenergic neurones. *Am. J. Physiol.* **125**, 765–777.

Cannon, W. B. & Rosenblueth, A. (1933). Sympathin-E and Sympathin-I. *Am. J. Physiol.* **104**, 557–574.

Cannon, W. B. & Uridil, J. E. (1921). Studies on the conditions of activity in endocrine glands. VIII. Some effects on the denervated heart of stimulating the nerves of the liver. *Am. J. Physiol.* **58**, 353–354.

Dixon, W. E. & Hamill, P. (1909). The mode of action of specific substances with special reference to secretin. *J. Physiol., Lond.* **38**, 314–336.

Elliott, T. R. (1905). The action of adrenaline. *J. Physiol., Lond.* **32**, 401–467.

Euler, U. S. von (1946*a*). The presence of a substance with sympathin-E properties in spleen extracts. *Acta physiol. scand.* **11**, 168–186.

Euler, U. S. von (1946*b*). The presence of a sympathomimetic substance in extracts of mammalian heart. *J. Physiol., Lond.* **105**, 38–44.

Euler, U. S. von (1946*c*). A specific sympathomimetic ergone in adrenergic nerve fibres (sympathin) and its relation to adrenaline and noradrenaline. *Acta physiol. scand.* **12**, 73–97.

Euler, U. S. von (1948). Identification of the sympathomimetic ergone in adrenergic nerves of cattle (sympathin-N) with laevo-noradrenaline. *Acta physiol. scand.* **16**, 63–74.

Euler, U. S. von (1950). Noradrenaline (arterenol), adrenal medullary hormone and chemical transmitter of adrenergic nerves. *Ergebn. Physiol.* **46**, 261–307.

Euler, U. S. von (1951). The nature of adrenergic nerve mediators. *Pharmac. Rev.* **3**, 247–277.

Euler, U. S. von (1956). *Noradrenaline*. Springfield, Illinois: Charles C. Thomas.

Euler, U. S. von (1958). Distribution and metabolism of catechol hormones in tissues and axones. *Recent Prog. Horm. Res.* **14**, 483–507.

Euler, U. S. von (1959). Autonomic neuroeffector transmission. In *Handbook of Physiology*, vol. I, Section I, Neurophysiology, p. 215. Washington, D.C.: American Physiology Society.

Euler, U. S. von & Lishajko, F. (1957). Dopamine in mammalian lung and spleen. *Acta physiol. pharmac. néerl.* **6**, 295–303.

Falck, B. (1962). Observations on the possibilities of the cellular localization of monoamines by a fluorescence method. *Acta physiol. scand.* **56**, Suppl. 197.

Falck, B., Häggendal, J. & Owman, C. (1963). The localization of adrenaline in adrenergic nerves in the frog. *Q. J. exp. Physiol.* **48**, 253–257.

Falck, B. & Torp, A. (1962). New evidence for the localization of noradrenaline in the adrenergic nerve terminals. *Med. exp.* **6**, 169–172.

Gaddum, J. H. & Goodwin, L. G. (1947). Experiments on liver sympathin. *J. Physiol., Lond.* **105**, 357–369.

Gaddum, J. H. & Kwiatkowski, H. (1939). Properties of the substance liberated by adrenergic nerves in the rabbit's ear. *J. Physiol., Lond.* **96**, 385–391.

Greer, C. M., Pinkston, J. O., Baxter, J. H. & Brannon, E. S. (1938). Norepinephrine (β-(3,4-dihydroxyphenyl)-β-hydroxyethylamine) as a possible mediator in the sympathetic division in the autonomic nervous system. *J. Pharmac. exp. Ther.* **62**, 189–227.

Holtz, P., Credner, K. & Kroneberg, G. (1947). Über das sympathicomimetische pressorische Prinzip des Harns ('Urosympathin'). *Arch. exp. Path. Pharmak.* **204**, 228–243.

Holtz, P., Heise, R. & Lüdtke, K. (1938). Fermentativer Abbau von *l*-dioxyphenylalanin (Dopa) durch Niere. *Arch. exp. Path. Pharmak.* **191**, 87–118.

Lissák, K. (1939). Effects of extracts of adrenergic fibres on the frog heart. *Am. J. Physiol.* **125**, 778–785.

Loewi, O. (1921). Über humorale Uebertragbarkeit der Herznervenwirkung. *Pflügers Arch. ges. Physiol.* **189**, 239–242.

Loewi, O. (1936). Quantitative und qualitative Untersuchungen über den Sympathicusstoff. *Pflügers Arch. ges. Physiol.* **237**, 504–514.

Mann, M. & West, G. B. (1950). The nature of hepatic and splenic sympathin. *Br. J. Pharmac. Chemother.* **5**, 173–177.

Mann, M. & West, G. B. (1951). The nature of uterine and intestinal sympathin. *Br. J. Pharmac. Chemother.* **6**, 79–82.

Oliver, G. & Schäfer, E. A. (1895). The physiological effects of extracts of the suprarenal capsules. *J. Physiol., Lond.* **18**, 230–276.

Outschoorn, A. S. (1952). The hormones of the adrenal medulla and their release. *Br. J. Pharmac. Chemother.* **7**, 605–615.

Outschoorn, A. S. & Vogt, M. (1952). The nature of cardiac sympathin in the dog. *Br. J. Pharmac. Chemother.* **7**, 319–324.

Peart, W. S. (1949). The nature of splenic sympathin. *J. Physiol., Lond.* **108**, 491–501.

Schmiterlöw, C. G. (1948). The nature and occurrence of pressor and depressor substances in extracts from blood vessels. *Acta physiol. scand.* **16**, Suppl. 56.

Stehle, R. L. & Ellsworth, H. C. (1937). Dihydroxyphenyl ethanolamine (arterenol) as a possible sympathetic hormone. *J. Pharmac. exp. Ther.* **59**, 114–121.

REFERENCES

Stolz, F. (1904). Über Adrenalin und Alkylaminoacetobrenzcatechin. *Ber. dtsch. chem. Ges.* **37**, 4149–4154.

Tainter, M. L. & Luduena, F. P. (1950). Sympathetic hormonal transmission. *Recent Prog. Horm. Res.* **5**, 3–35.

Takamine, J. (1901). The isolation of the active principle of the suprarenal gland. *J. Physiol., Lond.* **27**, 29–30*P*.

Tullar, B. F. (1948). The resolution of *dl*-arterenol. *J. Am. chem. Soc.* **70**, 2067–2068.

West, G. B. (1950). Further studies on sympathin. *Br. J. Pharmac. Chemother.* **5**, 165–172.

4

DISTRIBUTION OF CATECHOLAMINES AND THE STORAGE OF NORADRENALINE IN SYMPATHETIC NERVES

A. DISTRIBUTION OF CATECHOLAMINES

1. Distribution of noradrenaline and adrenaline

(*a*) *Peripheral nerves*

A systematic study of extracts of various mammalian nerves by Rexed & Euler (1951) revealed that there was a close correlation between the noradrenaline content and the proportion of non-myelinated autonomic nerve fibres, in keeping with the proposal that noradrenaline was the adrenergic transmitter. The finding that sympathetic ganglia contained about the same concentrations of noradrenaline as the postganglionic fibres suggested that the ganglion cells contain amounts of noradrenaline similar to those found in the axons.

The highest noradrenaline content was found in splenic nerve (Euler, 1956), which is in agreement with physiological and histological evidence indicating that this nerve contains almost exclusively small non-myelinated C fibres, and with the richness of the sympathetic innervation to this organ. The correlation between the noradrenaline content of a nerve and its content of adrenergic fibres is now so well established that the occurrence of noradrenaline in a nerve can be taken as evidence for the presence of adrenergic fibres. Thus the presence of noradrenaline in the vagus and in the short ciliary nerves suggested an unexpected adrenergic fibre component in these nerves (Euler, 1956).

(*b*) *Adrenal medulla*

Because of the relatively high concentrations of noradrenaline and adrenaline in the adrenal medulla their extraction and assay presents little technical difficulty. A large number of analyses using colorimetric, fluorimetric and bioassay procedures has been reported. The

Table 4.1. *Catecholamine content of some mammalian organs*

Organ or tissue	Species	Catecholamine content (μg/g) Nor-adrenaline	Adrenaline	Reference
Spleen	Cat	0·83	0·04	Rehn, 1958
Kidney	Cat	0·27	0·04	Rehn, 1958
Kidney	Ox	*c.* 0·07	—	Euler, 1956
Parotid gland	Rat	1·12	0·10	Strömblad & Nickerson, 1961
Submaxillary gland	Rat	1·04	0·05	Strömblad & Nickerson, 1961
Blood vessels	Ox	0·36	—	Euler & Lishajko, 1958
Skeletal muscle	Cat and rabbit	*c.* 0·10	Nil	Sedvall, 1963
Duodenum	Rat	0·30	—	Farrant *et al.* 1964
Brown fat (interscapular)	Mouse	0·49	0·05	Sidman *et al.* 1962
White fat (epididymal)	Mouse	0·05	0·01	Sidman *et al.* 1962
White fat (epididymal)	Rat	0·12	—	Paoletti *et al.* 1961
Lung	Ox	0·05	—	Euler, 1956
Ciliary body and iris	Rabbit	5·20	0·50	Eakins & Eakins, 1964
Lymph glands	Ox	*c.* 0·65	—	Euler, 1956
Placenta and bone marrow	Ox	Nil	Nil	Euler, 1956
Vas deferens	Guinea-pig	10·00	—	Sjöstrand, 1965
Heart	Rabbit	2·00	—	Goodall, 1951
Heart	Sheep	0·79	0·15	Goodall, 1951
Heart	Rat	0·96	< 0·05	Author's unpublished results

total catecholamine content is of the order of several milligrams per gram; the relative amounts of adrenaline and noradrenaline vary considerably among mammalian species (Euler, 1956).

(*c*) *Peripheral tissues*

Most peripheral tissues contain noradrenaline, though the concentration varies considerably, ranging from less than 0·1 μg/g in some skeletal muscles to more than 10 μg/g in the vas deferens of some species (Table 4.1). The ubiquitous distribution of noradrenaline is consistent with the presence of adrenergic vasomotor fibres in most peripheral tissues. The absence of noradrenaline in placenta or bone marrow is in agreement with physiological evidence which suggests that these tissues lack an adrenergic vasomotor innervation.

The relatively small amounts of noradrenaline in skeletal muscle (Sedvall, 1963) suggests that it is all present in the vasomotor fibres of small blood vessels in this tissue. However, the relatively high noradrenaline content of certain smooth muscle and glandular tissues, such as the vas deferens, ciliary body, iris and salivary glands, points to an adrenergic innervation of other than purely vasomotor function. In this context, the relatively high noradrenaline content of certain adipose tissues has been taken to indicate a possible adrenergic nervous control of lipid metabolism in such tissues (Sidman, Perkins & Weiner, 1962; Paoletti, Smith, Maickel & Brodie, 1961).

Studies of the noradrenaline content and the distribution of small nerve fibres in various layers of the walls of arteries and veins showed a clear correlation between the noradrenaline content and the adrenergic innervation of the various layers (Schmiterlöw, 1948). The distribution of noradrenaline in different regions of the mammalian heart has also been examined by several authors (Shore, Cohn, Higman & Maling, 1958; Muscholl, 1959*a*; Campos, Stitzel & Shideman, 1963; Klouda, 1963; Angelakos, 1965). In general the atria contain higher concentrations of noradrenaline than the ventricles and the right auricle contains more noradrenaline than the left auricle. These findings are in agreement with the known distribution of adrenergic fibres in the heart. Denervation studies have shown that the majority of the sympathetic innervation to both sides of the heart passes through the right stellate ganglion, though some fibres come from the left stellate ganglion to innervate the left auricle and ventricle (Goodall, 1951; Hertting & Schiefthaler, 1964).

The noradrenaline content of a peripheral tissue is remarkably constant under normal conditions. Prolonged stimulation of sympathetic nerves does not decrease the noradrenaline content of the nerve trunk (Luco & Goni, 1948) or of the innervated organ (Euler & Björkman, 1955). A diurnal rhythm for the noradrenaline content of rat liver was described by Hökfelt (1951) and a seasonal rhythm for the catecholamine content of various rat tissues was reported by Montagu (1959).

A diurnal rhythm for the noradrenaline content of rat pineal glands and salivary glands has recently been described (Wurtman & Axelrod, 1966).

The adrenaline content of peripheral tissues is more variable than the noradrenaline content. Most tissues contain small amounts of adrenaline, corresponding to from 2 to 15 % of the total catecholamine content (Table 4.1). In certain tissues adrenaline is contained in histologically distinct groups of chromaffin cells, and these undoubtedly account for the high adrenaline content of the abdominal aortic region (Biedl & Weisel, 1902; Muscholl & Vogt, 1964) and the Organ of Zückerkandl. Muscholl & Vogt (1958) demonstrated the presence of quite large amounts of adrenaline in the sympathetic ganglia of cats and rabbits; these ganglia also appear to contain chromaffin cells. The existence of scattered adrenaline-containing chromaffin cells has also been suggested in ovary, heart, salivary glands, spleen and kidney (Euler, 1963).

(*d*) *Central nervous system*

Noradrenaline is present in the mammalian brain and spinal cord (Euler, 1946; Holtz, 1950; McGeer & McGeer, 1962), though only negligible amounts of adrenaline are found in these tissues. A detailed survey of the distribution of noradrenaline in different regions of the brain revealed the presence of high concentrations of noradrenaline in certain basal regions, particularly in parts of the hypothalamus, where the concentration of noradrenaline may be as high as 2 μg/g (Vogt, 1954; Bertler & Rosengren, 1959). The specific regional localization of noradrenaline in the brain rules out the possibility that the catecholamine is present only in the adrenergic vasomotor innervation of brain blood vessels (Vogt, 1957). The possibility that noradrenaline may play a role in adrenergic transmission within the central nervous system has aroused great interest (see chapter 10). This hypothesis has been strengthened considerably by recent histochemical findings which have revealed a complex pattern of catecholamine-containing neurones and fibre tracts in the brain (Carlsson, Falck & Hillarp, 1962; Dahlström & Fuxe, 1964*a*; Fuxe, 1965). The distribution of catecholamines in various brain regions is illustrated in Table 4.2.

Table 4.2. *Distribution of catecholamines in the central nervous system*

Brain region	Species	Catecholamine content (μg/g)		Reference
		Noradrenaline	Dopamine	
Whole brain	Dog	0·16	0·19	Bertler & Rosengren, 1959
Whole brain	Rabbit	0·29	0·32	
Whole brain	Rat	0·49	0·60	
Whole brain	Cat	0·22	0·28	
Medulla oblongata	Dog	0·37	0·13	
Medulla oblongata	Rat	0·72	—	Glowinski & Iversen, 1966
Mesencephalon	Dog	0·33	0·20	Bertler & Rosengren, 1959
Pons	Dog	0·41	0·10	Bertler & Rosengren, 1959
Cerebellum	Rat	0·17	—	Glowinski & Iversen, 1966
Cerebellar cortex	Dog	0·07	—	Vogt, 1954
Cerebellar cortex	Dog	0·06	0·03	Bertler & Rosengren, 1959
Hypothalamus	Rat	1·79	—	Glowinski & Iversen, 1966
Hypothalamus	Rat	1·29	0·14	Laverty & Sharman, 1965
Hypothalamus	Dog	1·00	—	Vogt, 1954
Hypothalamus	Cat	1·40	—	Vogt, 1954
Striatum	Rat	0·25	7·50	Glowinski & Iversen, 1966
Caudate nucleus	Rat	0·27	6·39	Laverty & Sharman, 1965
Caudate nucleus	Cat	0·10	9·90	Laverty & Sharman, 1965
Lentiform nucleus	Dog	0·08	1·63	Bertler & Rosengren, 1959
Caudate nucleus	Dog	0·10	5·90	Bertler & Rosengren, 1959
Cerebral cortex	Dog	0·05	—	Vogt, 1954
Cerebral cortex	Rat	0·18	< 0·01	Laverty & Sharman, 1965
Cerebral cortex	Rat	0·24	—	Glowinski & Iversen, 1966

(*e*) *Body fluids*

The noradrenaline content of normal blood plasma is less than 1 μg/l. and the adrenaline content is barely detectable. Applications of the fluorimetric assay methods to measurements of plasma levels of noradrenaline and adrenaline have presented great technical difficulties because the methods must be used at their maximum limits of sensitivity. Recent estimates of normal plasma levels of noradrenaline are in the range 0·20–0·58 μg/l., and estimates of adrenaline levels range from zero to 0·22 μg/l. (Price & Price, 1957; Cohen & Goldenberg, 1957; Vendsalu, 1960; Weil-Malherbe, 1961; Häggendal, 1963).

The presence of noradrenaline in urine was first demonstrated by Holtz, Credner & Kroneberg (1947). The normal excretion of free noradrenaline and adrenaline in human urine ranges from 5 to 25 μg/24 h and from 1·3 to 5·4 μg/24 h respectively (Iisalo, 1962). From 1·5 to 3·0 times these amounts are present in the urine as conjugated forms of noradrenaline and adrenaline (Euler, 1956). Free and conjugated catecholamines, however, represent only 2–3 % of the total daily excretion of catecholamine metabolites in human or animal urine (Axelrod, Senoh & Witkop, 1958; Georges & Whitby, 1964).

(*f*) *Occurrence of noradrenaline and adrenaline in non-mammalian organisms*

Noradrenaline and adrenaline have been found to occur not only in the adrenal glands but also in other organs and secretions of various animal classes (Table 4.3). Catecholamines have been demonstrated in annelids, insects, elasmobranch and teleost fishes, amphibians and birds. The heart, brain and certain other tissues of the frog (*Rana pipiens*) and many other amphibians contain almost exclusively adrenaline with only traces of noradrenaline. The hearts of certain cyclostomes, such as the hagfish and lamprey, contain very high concentrations of adrenaline and noradrenaline—as much as 130 μg adrenaline/g and up to 50 μg noradrenaline/g (Östlund *et al.* 1960; Bloom *et al.* 1961; Hirsch, Jellinek & Cooper, 1964). Large amounts of adrenaline and noradrenaline are also found in some fruits and vegetables (Udenfriend, Lovenberg & Sjoerdsma, 1959).

Table 4.3. *Occurrence of catecholamines in non-mammalian organisms*

Reference	Species	Organ	Catecholamine content (μg/g tissue) Nor-adrenaline	Adrenaline
Bogdanski *et al.* 1963	Lizard	Heart	6·4	—
		Brain	1·6	—
	Frog, *R. pipiens*	Heart	—	1·2
		Brain	—	2·1
	Toad, *B. americanus*	Heart	—	1·8
		Brain	—	0·8
Östlund, 1954	Insects, various spp.	Whole	0·04–2·2	0·01–0·30
	Salmon	Heart	0·03	0·04
	Lumbricus	Ganglionic chain	0·32	1·40
Östlund *et al.* 1960	Lamprey, *Petromyzon*	Auricle	7	130
		Ventricle	—	30
	Hagfish, *Myxine*	Auricle	47	13
		Ventricle	6	45
Waalkes *et al.* 1958	Banana	Pulp	1·9	—
		Peel	122	—

2. Occurrence of dopamine and other related amines

Goodall (1951) detected the presence of dopamine in extracts of sheep adrenal glands and heart, and Euler, Hamberg & Hellner (1951) demonstrated the presence of dopamine in normal human urine. The development of a sensitive fluorimetric method for the assay of dopamine (Carlsson & Waldeck, 1958) was followed by reports of the presence of relatively large amounts of dopamine in several mammalian tissues. The presence of dopamine in adrenal extracts, where it constitutes some 2 % of the total catecholamines, was reported by Dengler (1957). Schümann (1956, 1958*a*, 1959) found that dopamine was present in many tissue extracts. In the ox, sheep and dog dopamine accounted for 50 % of the total catecholamine content of pancreas, parotid gland, spleen, brain and splenic nerves. Dopamine accounted for as much as 94–99 % of the total catecholamine content of certain other tissues such as lung, liver and intestine. In these tissues the dopamine content was particularly high in ruminant animals such as ox, sheep and goat. The dopamine content of sheep lung, for example, was found to be 24·0 μg/g, whereas that of dog lung was only 0·3 μg/g. The presence of dopamine in certain peripheral tissues was confirmed by Holtz (1959) and Wegmann (1963).

Dopamine also represents more than 50 % of the total catecholamine content in the brain and spinal cord of many species (Montagu, 1957, 1963; Carlsson, Lindquist, Magnusson & Waldeck, 1958; McGeer & McGeer, 1962). In the brain dopamine is specifically localized in certain basal nuclei, being particularly abundant in the striatum, which contains only small amounts of noradrenaline (Table 4.2) (Bertler & Rosengren, 1959).

Bertler *et al.* (1959) concluded that the large amounts of dopamine in certain peripheral tissues, especially in ruminant animals, was correlated with the presence of a special type of mast cell. This conclusion has been confirmed by fluorescence histochemical studies (Falck, Nystedt, Rosengren & Stenflo, 1964). The dopamine-containing cells occur in groups in the liver, duodenum, heart and lungs of ruminant animals and probably contain most of the dopamine in these tissues. The presence of dopamine-containing mast cells was also demonstrated in bovine splenic nerve trunks, which probably accounts for the high dopamine content of this tissue. It seems probable that most of the dopamine in peripheral tissues is present in such specialized mast cells rather than in adrenergic nerves as a precursor for noradrenaline synthesis, as was originally assumed (Schümann, 1956, 1958*a*). The function of the dopamine-rich mast cells remains obscure.

On the other hand, dopamine appears to have a specific distribution in the brain and spinal cord of mammals. It is present in relatively constant amounts in the brains of both ruminant and non-ruminant species, and is especially abundant in brain regions which are connected with extrapyramidal motor functions. The dopamine in these brain areas is present in nerve terminals and may play an independent role as a neurotransmitter substance in the central nervous system apart from its role as a precursor for noradrenaline synthesis (Carlsson *et al.* 1962; Fuxe, Hökfelt & Nilsson, 1964).

Small amounts of DOPA have been detected in sheep adrenals

(Goodall, 1951) and in human skin (Foster & Brown, 1957) where it is presumed to act as a precursor for melanin synthesis. DOPA is not present in detectable amounts in most peripheral tissues (Wegmann, 1963), although a specific and sensitive fluorimetric assay procedure is now available for its estimation.

Isoprenaline was reported to be present in trace amounts in mammalian adrenal extracts and in pulmonary venous blood (Lockett, 1954, 1957), but subsequent investigations failed to detect the presence of this compound in adrenal or heart extracts (Muscholl, 1959*b*).

Octopamine (*p*-hydroxyphenylethanolamine; norsynephrine) has been identified in extracts of the posterior salivary glands of the octopus (Erspamer & Boretti, 1951). Octopamine and other monophenolic amines, such as *p*- and *m*-tyramine and *p*-sympatol (synephrine) are excreted normally in human urine (review, see Axelrod, 1959). As much as 200–400 μg of tyramine are excreted daily (Sjoerdsma *et al.* 1959). It has been reported that tyramine is normally present in appreciable amounts in the brain and spinal cord of the rat, rabbit and dog (Spector, Melmon, Lovenberg & Sjoerdsma, 1963). However, Gunne & Jonsson (1965) failed to detect the presence of tyramine after ion-exchange chromatography of rabbit brain extracts. After the prolonged administration of inhibitors of monoamine oxidase, substantial accumulations of octopamine, β-phenylethylamine and other related amines may occur in mammalian tissues (Kakimoto & Armstrong, 1962; Nakajima, Kakimoto & Sano, 1964; Kopin, Fischer, Musacchio & Horst, 1964).

B. INTRACELLULAR STORAGE OF NORADRENALINE IN SYMPATHETIC NERVES

1. Localization of noradrenaline in sympathetic postganglionic nerves

It is impossible to isolate adrenergic nerve terminals from peripheral tissues, so that a direct study of the storage of noradrenaline in the terminal regions of these nerves has not been possible. On the other hand, there is strong evidence that the noradrenaline content of most peripheral tissues is entirely localized in sympathetic nerve fibres. If the postganglionic sympathetic nerves are severed and allowed to degenerate, the amount of noradrenaline in the denervated tissue falls to very low levels or disappears completely (Euler & Purkhold, 1951; Goodall, 1951). Such studies have been repeated many times and the use of surgical denervation is now a standard tool for the study of noradrenaline storage and inactivation in adrenergic nerves. Peripheral tissues thus do not themselves store noradrenaline, but contain the catecholamine only by virtue of their adrenergic innervation. After sympathetic denervation the noradrenaline content of sheep heart gradually reappears some 4–6 weeks later as the adrenergic innervation regenerates (Goodall, 1951). Noradrenaline does not occur in preganglionic sympathetic nerve fibres; sectioning of the preganglionic sympathetic nerves has little or no effect on the noradrenaline content of peripheral tissues or sympathetic ganglia (Rehn, 1958; Fischer & Snyder, 1965).

A comparison of the concentrations of noradrenaline in sympathetic ganglia and nerve trunks with the concentrations found in peripheral tissues shows that the amine must be highly concentrated in the terminal regions of the nerves. For example, ox spleen contains approximately 3 μg noradrenaline per gram wet weight and bovine splenic nerve contains approximately 15 μg/g. If the noradrenaline were evenly distributed throughout the axons this would imply that about 20 % of the wet weight of the spleen was accounted for by sympathetic nerve terminals. Since this is inconceivable, it follows that the transmitter must be present in much higher concentrations in the nerve terminals than in the preterminal axons of the nerve trunk (Euler, 1957). The conclusions drawn by this indirect approach have been verified in recent years by the fluorescence histochemical technique which allows a direct examination of the distribution of catecholamines in tissues (chapter 2). With this method it has been confirmed that the noradrenaline in peripheral tissues is exclusively localized in the terminals of postganglionic sympathetic nerves and that the transmitter is present in very high concentrations in these terminal regions (Falck & Torp, 1962). Although the fluorescence histochemical technique does not allow precise quantitative estimates to be made of the noradrenaline content of nerve terminals, Norberg & Hamberger (1964) have estimated the noradrenaline contents of sympathetic ganglion cells and preterminal axons to be 10–100 μg/g and 100–500 μg/g respectively. The concentration of noradrenaline in the intensely fluorescent sympathetic nerve terminals was estimated to be of the order of 10000 μg/g. It should be pointed out that the concentrations of noradrenaline in adrenergic cell bodies and preterminal axons are often too low to permit the histochemical demonstration of such structures unless their amine content is experimentally raised (Norberg & Hamberger, 1964; Dahlström & Fuxe, 1964*b*).

2. Anatomy of the autonomic innervation apparatus in peripheral tissues

Extensive investigations of central and peripheral adrenergic neurones have been pursued in recent years with the newly developed fluorescence histochemical technique. These studies have contributed greatly to our knowledge of the anatomy of central and peripheral adrenergic synapses. The considerable literature on the histochemical findings has been well reviewed by Norberg & Hamberger (1964), Dahlström & Fuxe (1964*a*), Malmfors (1965), and Falck & Owman (1966).

A description of the peripheral autonomic innervation in terms of nerve endings is misleading; a more accurate description is that of Hillarp (1946, 1959) who suggested the term 'ground plexus' to describe the autonomic innervation apparatus. In peripheral tissues free autonomic nerve endings are rarely if ever seen; the innervation consists of a dense network of very fine nerve fibres, which derive from the autonomic nerve trunk fibres by repeated branching (Plates 2, 3). The very fine fibres run in a surrounding plexus of Schwann-cell cytoplasm which anastomoses to form a network of fine strands. Within this plexus the autonomic fibres are present often in bundles of 2–4 axons, and run for considerable distances through the network, synapsing with many effector cells *en passage*. Thus, each adrenergic fibre probably innervates many effector cells and conversely each effector cell is innervated by many adrenergic fibres, which may arise from different neurones. Such an anatomical arrangement can explain the well-known phenomena of spatial and temporal summation in the adrenergic innervation apparatus, which were described by Cannon & Rosenblueth (1937) and Rosenblueth (1950). The earlier theories of these authors, and other theories of the anatomical arrange-

2 Fluorescence micrograph of rat iris 48 h after partial denervation of the sympathetic nerve supply. Under these conditions the profuse branching of individual fibres can be clearly seen. In this view at least two systems of branching terminals are seen, together with numerous preterminal axons. Magnification $\times 160$. From Malmfors (1965).

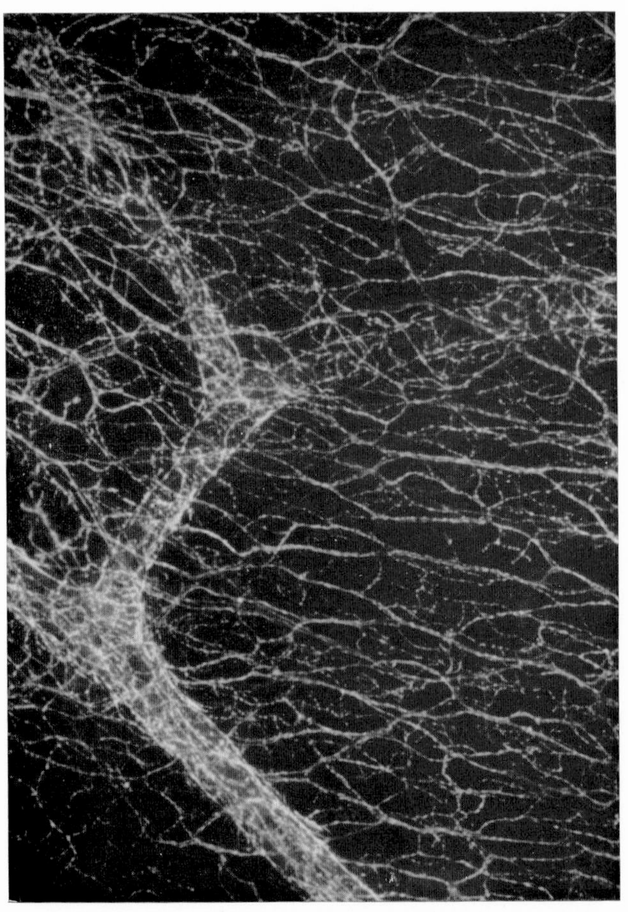

3 Fluorescence micrograph of normal rat iris. In addition to the strongly fluorescent and uniformly distributed plexus of varicose adrenergic nerve terminals over the dilator muscle, there is seen an arteriole surrounded by a similar adrenergic ground plexus (cf. Plate 2). Magnification ×160. From Malmfors (1965).

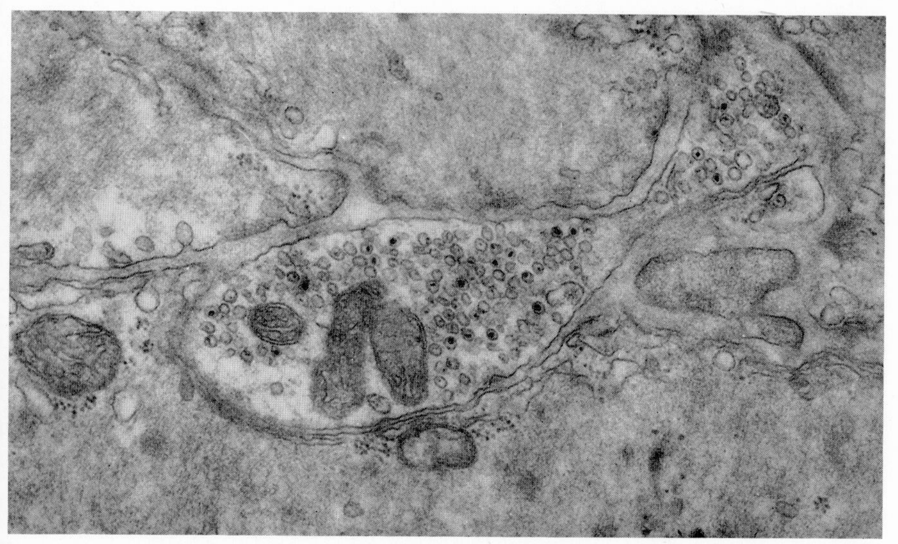

4 Electron micrograph showing nerve ending in the smooth muscle of the rat vas deferens. The section passes through what may be two adjacent varicosities of the adrenergic ground plexus. The cytoplasm within the nerve terminal contains mitochondria and numerous vesicles, many of which contain a densely stained core. Magnification ×38000. From K. C. Richardson, unpublished.

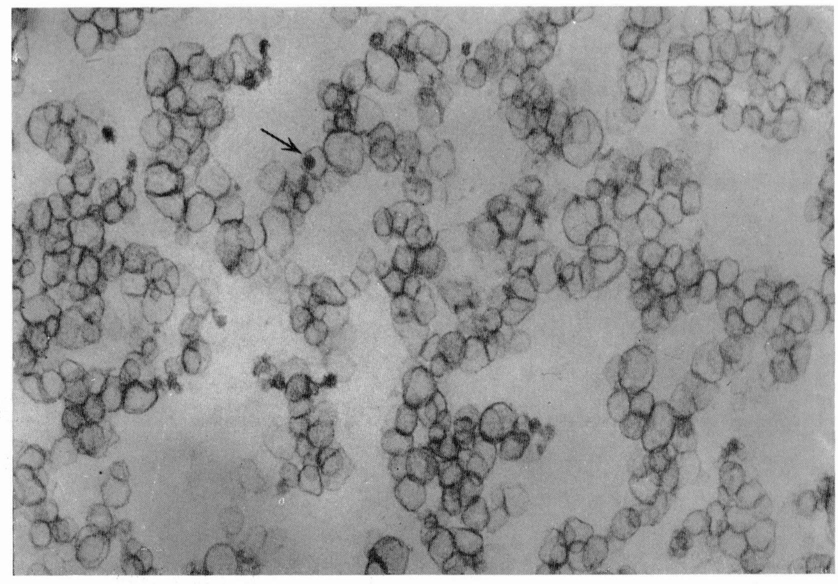

5 Electron micrograph of highly purified preparation containing noradrenaline storage particles from the rat heart. The fraction contains a homogeneous population of vesicles, some of which (indicated by an arrow) contain a densely stained core. The dense core vesicle indicated has a diameter of approximately 500 Å. Magnification × 50 000. From Potter (1966).

ment in terms of a syncytial nerve net which was in direct contact with the effector cells and which was in turn innervated by the autonomic nerve fibres themselves (i.e. the third link hypothesis, Boeke, 1949), have now generally been abandoned.

The fluorescence histochemical findings have demonstrated that noradrenaline is present in high concentrations throughout the length of the terminal fibres, which supports the view that each fibre may innervate several effector cells which it contacts along its course. Similar conclusions were reached by Richardson (1964) from an electron microscopic study of the anatomy of the sympathetic innervation of the rat iris. The histochemical evidence has also shown that the noradrenaline in terminal sympathetic fibres is not homogeneously distributed throughout the axoplasm but is concentrated at various thickenings along the fibre, giving the terminal fibres a characteristic varicose appearance (Plates 20, 30). This varicose appearance had been noted previously with conventional histological techniques (Garven & Gairns, 1952; Hillarp, 1959). These regions of high noradrenaline concentration probably correspond to the synaptic regions of the adrenergic terminals, although no definite synaptic membrane thickenings or contacts are apparent. The varicosities also contain dense groups of small membrane-bound particles, as illustrated in Plate 4 (Lever & Esterhuizen, 1961; De Robertis & De Iraldi, 1961; Richardson, 1962, 1964; Burnstock & Merrillees, 1964). In electron micrographs these particles are seen to be distributed in the axoplasm of the nerve fibres together with other intracellular particles such as mitochondria. The small particles are 400–500 Å in diameter and characteristically contain a deeply staining osmophilic core, though agranular particles are also found in adrenergic terminals (Plate 4).

3. Storage of noradrenaline in intracellular particles in adrenergic nerves

In 1953 Blaschko & Welch and Hillarp, Lagerstedt & Nilson demonstrated that within the cells of the adrenal medulla catecholamines are stored in dense granules which can be separated from mitochondria and other intracellular particles by density-gradient centrifugation. There have been many subsequent studies of the properties of these adrenal medullary storage particles. The concentrations of noradrenaline and adrenaline in the particles are so high that it seems unlikely that the amines could be present in free solution. The concentration of catecholamines in the granules has been estimated to be 0·6 M—more than twice the osmolarity of mammalian body fluids. It has therefore been suggested that the catecholamines are present within the granules in the form of a storage complex with other molecules. An adenine nucleotide–catecholamine complex appears to be the most likely possibility, since granules from the adrenal medulla of various species have been found to contain adenine nucleotides (usually adenosine triphosphate) and catecholamines in a constant ratio of 4 molecules of catecholamine to 1 molecule of adenine nucleotide. The ATP content of medullary granules is very high, amounting to a concentration of approximately 0·2 M within the granules (Hagen & Barrnett, 1960). The wet weight composition of adrenal medullary granules is as follows: water 68·5 %; catecholamines 6·7 %; adenine nucleotides 4·5 %; protein 11·5 %; lipids 7 %; other components 0·8 % (Hillarp, 1960*a*). Catecholamines are released from the granules by exposure to hypotonic solutions, mild acids, detergents, freezing and thawing, phospholipase action and heat, suggesting that the catecholamine complexes are retained within the granules by the outer membranous covering. Various alternative types of chemical complex have been suggested to explain the storage of catecholamines in adrenal granules (Green, 1962). Although there is strong circumstantial evidence in favour of the involvement of adenine nucleotides, there is as yet little indication of the nature of the complex. Recently Weiner & Jardetzky (1964) have presented evidence in favour of the feasibility of such a complex. The addition of ATP to an adrenaline solution resulted in a change in the nuclear magnetic resonance spectrum of adrenaline, indicating a stabilization of the molecule across the hydroxyl and amino groups of the side chain. It seems probable, however, that some other component, probably of a protein nature, must be involved *in vivo*.

There is also considerable evidence that noradrenaline is stored within particles in the terminals of sympathetic nerves. A considerable proportion of the total noradrenaline content of peripheral tissues can be isolated from tissue homogenates in a fraction of small particles on centrifugation (Potter & Axelrod, 1962; Campos &

Shideman, 1962; Euler & Lishajko, 1965). Particle-bound noradrenaline was first demonstrated in peripheral sympathetic nerves by Euler & Hillarp (1956) in homogenates of bovine splenic nerve. Euler and his co-workers have conducted an extensive investigation of the noradrenaline storage particles which can be isolated in a relatively pure form from splenic nerve homogenates or press juice (Euler, 1962). Noradrenaline storage particles from the terminals of sympathetic nerves have been isolated in highly purified preparations from homogenates of rat heart by Michaelson, Richardson, Snyder & Titus (1964), Snyder, Michaelson & Musacchio (1964) and Potter (1966) (Plate 5). The adrenergic nerve storage particles are similar in many respects to those isolated from the adrenal medulla. Within the nerve particles noradrenaline may be complexed with adenine nucleotides, which are present in the same stoichiometric amounts as in the adrenal medullary particles (Schümann, 1958*b*; Euler, Lishajko & Stjärne, 1963; Stjärne, 1964). Noradrenaline is spontaneously released from splenic nerve particles incubated at 30 °C and the release is accelerated by acids, detergents and hypotonic media (Euler & Lishajko, 1961, 1963). However, the adrenergic nerve storage particles differ in several important respects from the adrenal medullary storage particles; for an excellent review of these differences see Stjärne (1964). There are marked differences in size and density: the adrenal medullary particles may be as much as 1 μ in diameter and sediment in a sucrose density gradient in a layer of approximately 1·8 M sucrose, below mitochondria; adrenergic nerve particles are approximately 450 Å in diameter and are found in a layer of approximately 0·3 M sucrose on density gradients, well above the level of mitochondria. Furthermore the noradrenaline in nerve particles seems to be more resistant to release by detergents, freezing and thawing or hypotonicity than the catecholamines in adrenal medullary particles. These differences are sufficiently great to suggest that caution should be used in extrapolating the data obtained from studies of adrenal medullary particles to the storage of noradrenaline in peripheral sympathetic nerves.

It is not clear what proportion of the total noradrenaline content of sympathetic nerves is stored in intraneuronal particles. Whereas almost all of the catecholamines in adrenal medullary homogenates can be sedimented in a particle-bound form, studies on peripheral tissues have always revealed the presence of a substantial proportion of the total noradrenaline content in the soluble supernatant fraction after ultracentrifugation. Whether this amount of noradrenaline is actually present within sympathetic nerves in a free form in the axoplasm or whether the presence of noradrenaline in the supernatant merely reflects a technical artifact arising from the destruction of storage granules during the harsh disruption procedures used in these experiments remains obscure. Nor is it clear whether all of the noradrenaline in the storage particles is present in the form of a storage complex with adenine nucleotide. In both adrenal medullary and adrenergic nerve particles the molar ratio of catecholamines to ATP or other adenine nucleotides is somewhat greater than 4:1 (Hillarp, 1960*b*; Euler *et al.* 1963; Schumann, 1958*b*; Potter & Axelrod, 1963), suggesting that the particles may contain more than one pool of stored amines.

REFERENCES

Angelakos, E. (1965). Regional distribution of catecholamines in the dog heart. *Circulation Res.* **16**, 39–44.

Axelrod, J. (1959). Metabolism of norepinephrine and other sympathomimetic amines. *Physiol. Rev.* **39**, 751–776.

Axelrod, J., Senoh, J. S. & Witkop, B. B. (1958). *O*-Methylation of catecholamines *in vivo*. *J. biol. Chem.* **233**, 697–701.

Bertler, A., Falck, B., Hillarp, N.-Å., Rosengren, E. & Torp, A. (1959). Dopamine and chromaffin cells. *Acta physiol. scand.* **47**, 251–258.

Bertler, A. & Rosengren, A. (1959). Occurrence and distribution of catecholamines in brain. *Acta physiol. scand.* **47**, 350–361.

Biedl, A. & Wiesel, J. (1902). Über die funktionelle Bedeutung der Nebenorgane des Sympathicus (Zuckerkandl) und der chromaffine Zellgruppen. *Pflügers Arch. ges. Physiol.* **91**, 434–461.

Blaschko, H. & Welch, A. D. (1953). Localization of adrenaline in cytoplasmic particles of the bovine adrenal medulla. *Arch. exp. Path. Pharmak.* **219**, 17–22.

Bloom, G., Östlund, E., Euler, U. S. von, Lishajko, F., Ritzen, M. & Adams-Ray, J. (1961). Studies on catecholamine-containing granules of specific cells in cyclostome hearts. *Acta physiol. scand.* **53**, Suppl. 185.

Boeke, J. (1949). The sympathetic endformation, its synaptology, the interstitial cells, the periterminal network and its bearing on the neurone theory. Discussion and Critique. *Acta Anat.* **8**, 18–61.

Bogdanski, D. F., Bonomi, L. & Brodie, B. B. (1963). Occurrence of serotonin and catecholamines in brain and peripheral organs of various vertebrate classes. *Life Sciences* **2**, 80–84.

Burnstock, G. & Merrillees, N. C. R. (1964). Structural and experimental studies on autonomic nerve endings in smooth muscle. In *Pharmacology of Smooth Muscle* (Second International Pharmacol. Meeting, Prague, 1963). Oxford: Pergamon Press.

Campos, H. A. & Shideman, F. E. (1962). Subcellular distribution of catecholamines in the dog heart. Effects of reserpine and norepinephrine administration. *Int. J. Neuropharmac.* **1**, 13–22.

Campos, H. A., Stitzel, R. E. & Shideman, F. E. (1963). Action of tyramine and cocaine on catecholamine levels in subcellular fractions of the isolated cat heart. *J. Pharmac. exp. Ther.* **141**, 290–300.

Cannon, W. B. & Rosenblueth, A. (1937). *Autonomic Neuroeffector Systems.* New York: Macmillan.

Carlsson, A., Falck, B. & Hillarp, N.-Å. (1962). Cellular localization of brain monoamines. *Acta physiol. scand.* **56**, Suppl. 196.

Carlsson, A., Lindqvist, M., Magnusson, T. & Waldeck, B. (1958). On the presence of 3-hydroxytyramine in brain. *Science, N.Y.* **127**, 417.

Carlsson, A. & Waldeck, B. (1958). A fluorimetric method for the determination of dopamine (3-hydroxytyramine). *Acta physiol. scand.* **44**, 293–298.

Cohen, G. & Goldenberg, M. (1957). The simultaneous fluorimetric determination of adrenaline and noradrenaline in plasma. *J. Neurochem.* **2**, 58–70.

Dahlström, A. & Fuxe, K. (1964*a*). Evidence for the existence of monoamine-containing neurones in the central nervous system. I. Demonstration of monoamines in the cell bodies of brain stem neurones. *Acta physiol. scand.* **62**, Suppl. 232.

Dahlström, A. & Fuxe, K. (1964*b*). A method for the demonstration of adrenergic nerve fibres in peripheral nerves. *Z. Zellforsch. mikrosk. Anat.* **62**, 602–607.

Dengler, H. (1957). Über das Vorkommen von Oxytyramin in der Nebenniere. *Arch. exp. Path. Pharmak.* **231**, 373–377.

De Robertis, E. & De Iraldi, A. P. (1961). Plurivesicular secretory processes and nerve endings in the pineal gland of the rat. *J. biophys. biochem. Cytol.* **10**, 361–372.

Eakins, K. E. & Eakins, H. M. T. (1964). Adrenergic mechanisms and the outflow of aqueous humor from the rabbit eye. *J. Pharmac. exp. Ther.* **144**, 60–65.

Erspamer, V. & Boretti, G. (1951). Identification and characterization by paper chromatography of enteramine, octopamine, tyramine, histamine and allied substances in extracts of posterior salivary glands of Octopoda and in other tissue extracts of Vertebrates and Invertebrates. *Archs. int. Pharmacodyn. Thér.* **88**, 296–332.

Euler, U. S. von (1946). A specific sympathomimetic ergone in adrenergic nerve fibres (sympathin) and its relation to adrenaline and nor-adrenaline. *Acta physiol. scand.* **12**, 73–97.

Euler, U. S. von (1956). *Noradrenaline.* Springfield, Illinois: Charles C. Thomas.

Euler, U. S. von (1957). In *Metabolism of the Nervous System* (ed. D. Richter), p. 543, Oxford: Pergamon Press.

Euler, U. S. von (1959). Autonomic neuroeffector transmission. In *Handbook of Physiology*, vol. I, Section I, Neurophysiology, p. 215. Washington, D.C.: American Physiology Society.

Euler, U. S. von (1962). In *Neurochemistry* (ed. Elliott, Page and Quastel), p. 578. Springfield, Illinois: Charles C. Thomas.

Euler, U. S. von (1963). In *Comparative Endocrinology* (ed. Euler and Heller), vol. I, p. 258. New York: Academic Press.

Euler, U. S. von & Björkman, S. (1955). Effect of increased adrenergic nerve activity on the content of noradrenaline and adrenaline in cat organs. *Acta physiol. scand.* **33**, Suppl. 118, pp. 17–20.

Euler, U. S. von, Hamberg, U. & Hellner, S. (1951). *β*-(3:4-Dihydroxyphenyl)-ethylamine (hydroxytyramine) in normal human urine. *Biochem. J.* **49**, 655–658.

Euler, U. S. von & Hillarp, N.-Å. (1956). Evidence for the presence of noradrenaline in submicroscopic structures of adrenergic axones. *Nature, Lond.* **177**, 43–45.

Euler, U. S. von & Lishajko, F. (1958). Catecholamines in the vascular wall. *Acta physiol. scand.* **42**, 333–341.

Euler, U. S. von & Lishajko, F. (1961). Noradrenaline release from isolated nerve granules. *Acta physiol. scand.* **51**, 193–203.

Euler, U. S. von & Lishajko, F. (1963). Catecholamine release and uptake in isolated adrenergic nerve granules. *Acta physiol. scand.* **57**, 468–480.

Euler, U. S. von & Lishajko, F. (1965). Free and bound noradrenaline in the rabbit heart. *Nature, Lond.* **205**, 179–180.

Euler, U. S. von, Lishajko, F. & Stjärne, L. (1963). Catecholamines and adenosine triphosphate in isolated adrenergic nerve granules. *Acta physiol. scand.* **59**, 495–496.

Euler, U. S. von & Purkhold, A. (1951). Effect of sympathetic denervation on the noradrenaline and adrenaline content of the spleen, kidney and salivary glands in the sheep. *Acta physiol. scand.* **24**, 212–217.

Falk, B., Nystedt, T., Rosengren, E. & Stenflo, J. (1964). Dopamine and mast cells in ruminants. *Acta pharmac. tox.* **21**, 51–81.

Falck, B. & Owman, C. (1966). Histochemistry of monoamine mechanisms in peripheral neurones. In *Mechanisms of Release of Biogenic Amines.* Oxford: Pergamon Press.

Falck, B. & Torp, A. (1962). New evidence for the localization of noradrenaline in the adrenergic nerve terminals. *Med. exp.* **6**, 169–172.

Farrant, J., Harvey, J. A. & Pennefather, J. N. (1964). The influence of phenoxybenzamine on the storage of noradrenaline in rat and cat tissues. *Br. J. Pharmac. Chemother.* **22**, 104–112.

Fischer, J. E. & Snyder, S. H. (1965). Disposition of norepinephrine-H^3 in sympathetic ganglia. *J. Pharmac. exp. Ther.* **150**, 190–195.

Foster, M. & Brown, S. R. (1957). The production of dopa by normal pigmented mammalian skin. *J. biol. Chem.* **225**, 247–252.

Fuxe, K. (1965). Evidence for the existence of monoamine neurons in the central nervous system. III. The monoamine nerve terminal. *Z. Zellforsch. mikrosk. Anat.* **65**, 573–596.

Fuxe, K., Hökfelt, T. & Nilsson, O. (1964). Observations on the cellular localization of dopamine in the caudate nucleus of the rat. *Z. Zellforsch. mikrosk. Anat.* **63**, 701–706.

Garven, H. S. & Gairns, F. W. (1952). The silver diamine ion staining of peripheral nerve elements and the interpretation of the results: with a modification of the Bielschowsky–Gros method for frozen sections. *Q. J. exp. Physiol.* **37**, 131–142.

Georges, R. & Whitby, L. G. (1964). Urinary excretion of metabolites of catecholamines in normal individuals and hypertensive patients. *J. clin. Path.* **17**, 64–69.

Glowinski, J. & Iversen, L. L. (1966). Regional studies of catecholamines in the rat brain, I. *J. Neurochem.* **13**, 655–669.

Goodall, McC. (1951). Studies of adrenaline and noradrenaline in mammalian hearts and suprarenals. *Acta physiol. scand.* **24**, Suppl. 85.

Green, J. P. (1962). Binding of some biogenic amines in tissues. *Adv. Pharmac.* **1**, 349–422.

Gunne, L. M. & Jonsson, J. (1965). On the occurrence of tyramine in the rabbit brain. *Acta physiol. scand.* **64**, 434–438.

Hagen, P. & Barrnett, R. J. (1960). In *Adrenergic Mechanisms*, CIBA, p. 83. London: Churchill.

Häggendal, J. (1963). An improved method for fluorimetric determination of small amounts of adrenaline and noradrenaline in plasma and tissues. *Acta physiol. scand.* **59**, 242–254.

Hertting, G. & Schiefthaler, T. (1964). The effect of stellate ganglion excision on the catecholamine content and the uptake of H^3-norepinephrine in the heart of the cat. *Int. J. Neuropharmac.* **3**, 65–69.

Hillarp, N.-Å. (1946). Structure of the synapse and the peripheral innervation apparatus of the autonomic nervous system. *Acta anat.* **2**, Suppl. 4.

Hillarp, N.-Å. (1959). The construction and functional organization of the autonomic innervation apparatus. *Acta physiol. scand.* **46**, Suppl. 157.

Hillarp, N.-Å. (1960*a*). In *Adrenergic Mechanisms*, CIBA, p. 481. London: Churchill.

Hillarp, N.-Å. (1960*b*). Different pools of catecholamines stored in the adrenal medulla. *Acta physiol. scand.* **50**, 8–22.

Hillarp, N.-Å., Lagerstedt, S. & Nilson, B. (1953). The isolation of a granular fraction from the suprarenal medulla, containing the sympathomimetic catecholamines. *Acta physiol. scand.* **29**, 251–263.

Hirsch, E. F., Jellinek, M. & Cooper, T. (1964). Innervation of the systemic heart of the California hagfish. *Circulation Res.* **14**, 212–217.

Hökfelt, B. (1951). Noradrenaline and adrenaline in mammalian tissues. Distribution under normal and pathological conditions with special reference to the endocrine system. *Acta physiol. scand.* **25**, Suppl. 92.

Holtz, P. (1950). Über die sympathicomimetische Wirksamkeit von Gehirnextrakten. *Acta physiol. scand.* **20**, 354–362.

Holtz, P. (1959). Role of L-dopa decarboxylase in the biosynthesis of catecholamines in nervous tissue and the adrenal medulla. *Pharmac. Rev.* **11**, 317–329.

Holtz, P., Credner, K. & Kroneberg, G. (1947). Über das sympathicomimetische pressorische Prinzip des Harns ('Urosympathin'). *Arch. exp. Path. Pharmak.* **204**, 228–243.

Iisalo, E. (1962). Enzyme action on noradrenaline and adrenaline. *Acta pharmac. tox.* **19**, Suppl. 1.

Kakimoto, Y. & Armstrong, M. D. (1962). On the identification of octopamine in mammals. *J. biol. Chem.* **237**, 422–427.

Klouda, M. A. (1963). Distribution of catecholamines in the dog heart. *Proc. Soc. exp. Biol. Med.* **112**, 728–729.

Kopin, I. J., Fischer, J. E., Musacchio, J. & Horst, W. D. (1964). Evidence for a false neurochemical transmitter as a mechanism of the hypotensive effect of monoamine oxidase inhibitors. *Proc. natn. Acad. Sci. U.S.A.* **52**, 716–721.

Laverty, R. & Sharman, D. F. (1965). The estimation of small quantities of 3,4-dihydroxyphenylethylamine in tissues. *Br. J. Pharmac. Chemother.* **24**, 538–548.

Lever, J. D. & Esterhuizen, A. C. (1961). Fine structure of the arteriolar nerves in the guinea pig pancreas. *Nature, Lond.* **192**, 566–567.

Lockett, M. F. (1954). Identification of an isoprenaline-like substance in extracts of adrenal glands. *Br. J. Pharmac. Chemother.* **9**, 498–505.

Lockett, M. F. (1957). The transmitter released by stimulation of the bronchial sympathetic nerves of cats. *Br. J. Pharmac. Chemother.* **12**, 86–96.

Luco, J. V. & Goni, F. (1948). Synaptic fatigue and chemical mediators of post-ganglionic fibres. *J. Neurophysiol.* **11**, 497–500.

McGeer, E. G. & McGeer, P. L. (1962). Catecholamine content of spinal cord. *Can. J. Biochem. Physiol.* **40**, 1141–1151.

Malmfors, T. (1965). Studies on adrenergic nerves. *Acta physiol. scand.* **64**, Suppl. 248.

Michaelson, I. A., Richardson, K. C., Snyder, S. H. & Titus, E. O. (1964). The separation of catecholamine storage vesicles from rat heart. *Life Sciences* **3**, 971–978.

Montagu, K. A. (1957). Catechol compounds in rat tissues and in brains of different animals. *Nature, Lond.* **180**, 244–245.

Montagu, K. A. (1959). Seasonal changes of the catechol compounds present in rat tissues. *Biochem. J.* **71**, 91–99.

Montagu, K. A. (1963). Some catechol compounds other than noradrenaline and adrenaline in brains. *Biochem. J.* **86**, 9–11.

Muscholl, E. (1959*a*). Die Konzentration von Noradrenalin und Adrenalin in den einzelnen Abschnitten des Hertzens. *Arch. exp. Path. Pharmak.* **237**, 350–364.

Muscholl, E. (1959*b*). Zur Frage des Vorkommens von Isopropylnoradrenalin in Herz und Nebenniere. *Arch. exp. Path. Pharmak.* **237**, 365–374.

Muscholl, E. & Vogt, M. (1958). The action of reserpine on the peripheral sympathetic system. *J. Physiol., Lond.* **141**, 132–155.

Muscholl, E. & Vogt, M. (1964). Secretory responses of extramedullary chromaffin tissue. *Br. J. Pharmac. Chemother.* **22**, 193–203.

Nakajima, T., Kakimoto, Y. & Sano, I. (1964). Formation of β-phenylethylamine in mammalian tissue and its effect on motor activity in the mouse. *J. Pharmac. exp. Ther.* **143**, 319–325.

Norberg, K. A. & Hamberger, B. (1964). The sympathetic adrenergic neuron. *Acta physiol. scand.* **63**, Suppl. 238.

Östlund, E. (1954). The distribution of catechol amines in lower animals and their effect on the heart. *Acta physiol. scand.* **31**, Suppl. 112.

Östlund, E., Bloom, G., Adams-Ray, J., Ritzen, M., Siegran, M., Nordenstam, N., Lishajko, F. & Euler, U. S. von (1960). Storage and release of catecholamines and the occurrence of a specific submicroscopic granulation in hearts of Cyclostomes. *Nature, Lond.* **188**, 324–325.

Paoletti, R., Smith, R. L., Maickel, R. P. & Brodie, B. B. (1961). Identification and physiological role of norepinephrine in adipose tissue. *Biochem. biophys. Res. Commun.* **5**, 424–429.

Potter, L. T. (1966). Storage of norepinephrine in sympathetic nerves. Second International Catecholamine Symposium. *Pharmac. Rev.* **18,** 439–452.

Potter, L. T. & Axelrod, J. (1962). Intracellular localization of catecholamines in tissues of the rat. *Nature, Lond.* **194**, 581–582.

Potter, L. T. & Axelrod, J. (1963). Properties of norepinephrine storage particles of the rat heart. *J. Pharmac. exp. Ther.* **142**, 299–305.

Price, H. L. & Price, M. L. (1957). The chemical estimation of epinephrine and norepinephrine in human and canine plasma. *J. Lab. clin. Med.* **50,** 769–777.

Rehn, N. O. (1958). Effect of decentralization on the content of catechol amines in the spleen, and kidney of the cat. *Acta physiol. scand.* **42,** 309–312.

Rexed, C. & Euler, U. S. von (1951). The presence of histamine and noradrenaline in nerves as related to their content of myelinated and unmyelinated fibres. *Acta psychiat. neurol. scand.* **26,** 61–65.

Richardson, K. C. (1962). The fine structure of autonomic nerve endings in smooth muscle of the rat vas deferens. *J. Anat.* **96**, 427–442.

Richardson, K. C. (1964). The fine structure of the albino rabbit iris with special reference to the identification of adrenergic and cholinergic nerves and nerve endings in its intrinsic muscles. *Am. J. Anat.* **114,** 173–184.

Rosenblueth, A. (1950). *The Transmission of Nerve Impulses at Neuroeffector Junctions and Peripheral Synapses*. New York: John Wiley and Sons Inc.

Schmiterlöw, C. G. (1948). The nature and occurrence of pressor and depressor substances in extracts from blood vessels. *Acta physiol. scand.* **16,** Suppl. 56.

Schümann, H. J. (1956). Nachweis von Oxytyramin (Dopamin) in sympathischen Nerven und Ganglien. *Arch exp. Path. Pharmak.* **227**, 566–573.

Schümann, H. J. (1958*a*). Über die Verteilung von Noradrenalin und Hydroxytyramin in sympathischen Nerven (Milznerven). *Arch exp. Path. Pharmak.* **234,** 17–25.

Schümann, H. J. (1958*b*). Über den Noradrenalin und ATP Gehalt sympathischen Nerven. *Arch. exp. Path. Pharmak.* **233,** 296–300.

Schümann, H. J. (1959). Über den Hydroxytyramin Gehalt der Organe. *Arch exp. Path. Pharmak.* **236**, 474–482.

Sedvall, G. (1963). Noradrenaline storage in skeletal muscle. *Acta physiol. scand.* **60**, 39–50.

Shore, A., Cohn, V. H., Higman, B. & Maling, H. M. (1958). Distribution of norepinephrine in the heart. *Nature, Lond.* **181,** 848–849.

Sidman, R. L., Perkins, M. & Weiner, N. (1962). Noradrenaline and adrenaline content of adipose tissue. *Nature, Lond.* **193**, 36–37.

Sjoerdsma, A., Lovenberg, W., Oates, J. A., Crout, J. R. & Udenfriend, S. (1959). Alterations in the pattern of amine excretion in man produced by a monoamine oxidase inhibitor. *Science, N.Y.* **130**, 225.

Sjöstrand, N. O. (1965). The adrenergic innervation of the vas deferens and the accessory male genital glands. *Acta physiol. scand.* **65**, Suppl. 257.

Snyder, S. H., Michaelson, I. A. & Musacchio, J. (1964). Purification of norepinephrine storage granules from rat heart. *Life Sciences* **3**, 965–970.

Spector, S., Melmon, K., Lovenberg, W. & Sjoerdsma, A. (1963). The presence and distribution of tyramine in mammalian tissues. *J. Pharmac. exp. Ther.* **140,** 229–235.

Stjärne, L. (1964). Studies of catecholamine uptake, storage and release mechanisms. *Acta physiol. scand.* **62**, Suppl. 228.

Strömblad, B. C. R. & Nickerson, M. (1961). Accumulation of epinephrine and norepinephrine by some rat tissues. *J. Pharmac. ex. Ther.* **134**, 154–159.

Udenfriend, S., Lovenberg, W. & Sjoerdsma, A. (1959). Physiologically active amines in common fruits and vegetables. *Archs. Biochem. Biophys.* **85**, 487–490.

Vendsalu, A. (1960). Studies on adrenaline and noradrenaline in human plasma. *Acta physiol. scand.* **49**, Suppl. 173.

Vogt, M. (1954). The concentration of sympathin in different parts of the central nervous system under normal conditions and after administration of drugs. *J. Physiol., Lond.* **123**, 451–481.

Vogt, M. (1957). In *Metabolism of the Nervous System* (ed. D. Richter), p. 553. Oxford: Pergamon Press.

Waalkes, T. P., Sjoerdsma, A., Creveling, C. R., Weissbach, H. & Udenfriend, S. (1958). Serotonin, norepinephrine, and related compounds in bananas. *Science, N.Y.* **127**, 648–650.

Wegmann, A. (1963). Determination of 3-hydroxytyramine and dopa in various organs of the dog after dopa infusion. *Arch exp. Path. Pharmak.* **246**, 184–189.

Weil-Malherbe, H. (1961). In *Methods in Medical Research* (ed. Quastel), vol. IX, p. 130.

Weiner, N. & Jardetzky, O. (1964). A study of catecholamine nucleotide complexes by nuclear magnetic resonance spectroscopy. *Arch. exp. Path. Pharmak.* **248**, 308–318.

Wurtman, R. J. & Axelrod, J. (1966). A 24-hour rhythm in the content of norepinephrine in the pineal and salivary glands of the rat. *Life Sciences.* **5**, 665–669.

5

THE METABOLISM OF CATECHOLAMINES

A. BIOSYNTHESIS

1. Introduction

Shortly after the discovery of the enzyme L-DOPA decarboxylase by Holtz, Heise & Ludtke (1938), Blaschko (1939) proposed a hypothetical series of reactions by which adrenaline and noradrenaline might be formed from tyrosine *in vivo*. This is now known to be the main pathway for the biosynthesis of catecholamines in animal tissues:

L-tyrosine → L-DOPA → dopamine → L-noradrenaline → L-adrenaline.

The first direct evidence for this series of reactions came from studies of the synthesis of catecholamines in the adrenal medulla. It is now well established that adrenal medullary tissue, both *in vivo* and *in vitro*, is able to convert radioactively labelled tyrosine or DOPA into dopamine, noradrenaline and adrenaline (Demis, Blaschko & Welch, 1955; Hagen, 1956; Pellerin & D'Iorio, 1957; Kirshner & Goodall, 1956; Masuoka, Schott, Akawie & Clark, 1956; Udenfriend & Wyngaarden, 1956).

It has been technically more difficult to demonstrate noradrenaline synthesis in sympathetic nerves because of the small amounts of this substance present in peripheral tissues. It is only recently, with the advent of chromatographic techniques for the separation of noradrenaline from its precursors and metabolites, and with the availability of radioactively labelled precursors of high specific activity, that the biosynthesis of noradrenaline has been demonstrated in tissues other than the adrenal medulla.

In 1958 Goodall & Kirshner showed that homogenates of sympathetic nerves or ganglia were able to convert labelled tyrosine or DOPA into dopamine and noradrenaline. Labelled dopamine and noradrenaline were also formed when rat and cat brain slices were incubated with C^{14}-tyrosine (Masuoka, Clark & Schott, 1961; Masuoka, Schott & Petriello, 1963). Goldstein, Musacchio & Contrera (1962) found radioactive noradrenaline in the rabbit heart after the intravenous infusion of C^{14}-dopamine. More recently the synthesis of dopamine and noradrenaline has been demonstrated in isolated guinea-pig hearts perfused with labelled tyrosine (Spector *et al.* 1963); in dog hearts perfused with labelled dopamine (Chidsey, Kaiser & Braunwald, 1963); and in rabbit hearts perfused with labelled tyrosine or dopamine (Musacchio & Goldstein, 1963).

2. Enzymes involved in biosynthesis

The conversion of L-tyrosine to L-DOPA is catalysed by the enzyme tyrosine hydroxylase; the properties of this enzyme have been described in a recent review (Udenfriend, 1966). Tyrosine hydroxylase has been studied in brain, adrenal medulla and sympathetically innervated tissues by Nagatsu, Levitt & Udenfriend (1964) and in bovine caudate nucleus by Bagchi & McGeer (1964). Except in the adrenal medulla, the enzyme is largely associated with particles sedimenting at 15000 ***g***. Partially purified preparations from the supernatant fraction of adrenal medullary extracts require tetrahydropteridine cofactors and are stimulated by Fe^{2+}. Of the compounds tested, only L-tyrosine served as a substrate; D-tyrosine, tyramine, DL-*m*-tyrosine and L-tryptophan were inactive.

The decarboxylation of L-DOPA is well-documented (Holtz, 1959; Sourkes, 1966). A single enzyme appears to be responsible for the decarboxylation of L-DOPA, L-5-hydroxytryptophan and other aromatic L-amino acids, including *m*-tyrosine, *o*-tyrosine and possibly *p*-tyrosine, to form the corresponding aromatic amines (Hagen, 1962; Lovenberg, Weissbach & Udenfriend, 1962). The name 'Aromatic L-amino acid decarboxylase' would thus seem to be more appropriate than the commonly used 'DOPA decarboxylase'.

The decarboxylase is widely distributed in peripheral tissues and is also active in those regions of the brain known to contain high concentrations of catecholamines or serotonin (Holtz & Westermann, 1956). The considerable activity found in the kidney is noteworthy, since this may account for the large amounts of dopamine, tyramine and other aromatic amines found in normal human urine. In sympathetic nerves and adrenal medulla the enzyme is largely found in the supernatant fraction after the centrifugation of tissue homogenates

(Blaschko, Hagen & Welch, 1955; Schümann, 1958). Recently, however, Stjärne (1966) has reported that considerable decarboxylase activity is associated with washed suspensions of noradrenaline storage particles isolated from bovine splenic nerves. The decarboxylase has been purified by Fellman (1959) and by Werle & Aures (1959). In common with other amino acid decarboxylases, pyridoxal phosphate is required as a cofactor.

The final stage in the biosynthesis of noradrenaline involves the β-hydroxylation of dopamine, catalysed by the enzyme dopamine-β-hydroxylase. For a recent review of the properties of this enzyme see Kaufman & Friedman (1965). In the adrenal medulla and in peripheral tissues the enzyme is associated with particulate cellular components and it seems likely that the enzyme is located in storage granules in the adrenal medulla (Kirshner, 1959). Potter & Axelrod (1963) reported a similar association of dopamine-β-hydroxylase activity with catecholamine storage particles obtained from rat heart homogenates. The enzyme is distributed in various regions of the brain in amounts corresponding with the known distribution of catecholamines (Udenfriend & Creveling, 1959). The enzyme has been purified from ox adrenal glands by Levin & Kaufman (1961); it is a copper-containing protein which requires for activity the presence of ascorbic acid and oxygen (Friedman & Kaufman, 1965). In addition to dopamine, several other phenylethylamine derivatives can be β-hydroxylated by this enzyme. Among these alternative substrates are tyramine (converted to norsynephrine), phenylethylamine (to phenylethanolamine) and α-methyl dopamine (to α-methyl noradrenaline).

In the adrenal medulla the further conversion of noradrenaline to adrenaline is catalysed by the enzyme phenylethanolamine *N*-methyl transferase (Axelrod, 1962, 1966). This enzyme is found in the supernatant fraction of adrenal homogenates and in common with other methyl transferases requires the presence of *S*-adenosyl methionine as a methyl donor. It is doubtful whether the formation of adrenaline occurs to any significant extent in peripheral sympathetic nerves or in the brain, although Axelrod (1962) reported a slight *N*-methyl transferase activity in the rat heart and other authors have found traces of labelled adrenaline in brain after the administration of labelled precursors (Milhaud & Glowinski, 1962; McGeer & McGeer, 1964).

3. Alternative biosynthetic pathways

Because of the broad specificity of aromatic L-amino acid decarboxylase and dopamine-β-hydroxylase there are numerous possible alternative routes by which catecholamines might be synthesized from the precursors phenylalanine and tyrosine (Fig. 5.1). Several of these pathways have been demonstrated to be possible *in vivo* (indicated by

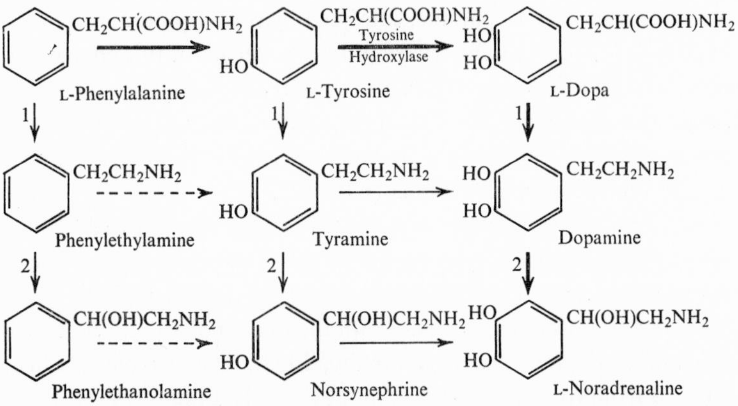

Fig. 5.1. Pathways for the biosynthesis of noradrenaline. Major pathway indicated by heavy arrows. Various alternative pathways are also possible as indicated by thin continuous arrows (reactions demonstrated to occur *in vivo*) and by dotted arrows (reactions known to be possible from *in vitro* studies, but not demonstrated *in vivo*). 1 = aromatic-L-amino acid decarboxylase; 2 = dopamine-β-hydroxylase.

the thin continuous lines in Fig. 5.1). Norsynephrine (octopamine) has been shown to occur in mammalian tissues and small amounts of tyramine may also be found (chapter 4). The formation of norsynephrine from tyramine has been demonstrated in the perfused rabbit heart by Musacchio & Goldstein (1963), and in the mouse heart after the intravenous injection of radioactively labelled tyramine by Carlsson & Waldeck (1963*a*). Although previous experiments had shown that tyramine does not act as a precursor for catecholamine synthesis in the adrenal medulla (Udenfriend & Wyngaarden, 1956), appreciable amounts of labelled noradrenaline and noradrenaline metabolites were found in the urine after the injection of labelled tyramine and norsynephrine in rats (Creveling, Levitt & Udenfriend, 1962). The amounts of injected tyramine or norsynephrine converted to noradrenaline in these experiments, however, were very small—ranging from 0·02 to 1·54 % of the injected doses. The conversion of tyramine to norsynephrine, on the other hand, occurs rapidly after the administration of tyramine (Carlsson & Waldeck, 1963*a*). Although it is doubtful whether these reactions have any physiological importance, they may be of considerable importance for an understanding of the pharmacological actions of tyramine and other phenylethylamine derivatives.

Dihydroxyphenylserine (DOPS) can be decarboxylated by aromatic L-amino acid decarboxylase both *in vitro* and *in vivo* (Blaschko, Burn & Langemann, 1950; Schmiterlöw, 1951) to form noradrenaline. Although this reaction probably does not occur under normal conditions it may provide a method for introducing noradrenaline into peripheral catecholamine stores (Carlsson, 1964). The use of DOPS or even L-DOPA itself as a precursor for catecholamine synthesis is complicated, however, by the widespread distribution and broad specificity of aromatic L-amino acid decarboxylase. Because of this, it is possible that the catecholamines formed by the decarboxylation of DOPS or DOPA may be present in wholly artificial cellular locations. After DOPS, for example, noradrenaline might be formed not only in cells which normally contain noradrenaline but also in cells which normally contain dopamine or serotonin.

α-Methylated amines derived from the decarboxylation of α-methyl DOPA and α-methyl-*m*-tyrosine can be further converted to the corresponding β-hydroxylated amines *in vivo* (α-methyl noradrenaline and metaraminol respectively) (Carlsson & Lindqvist, 1962; Gessa, Costa, Kuntzman & Brodie, 1962). After the administration of the α-methyl amino acids the α-methylated amines, α-methyl noradrenaline and metaraminol, have been shown to accumulate in brain and heart (Maître & Staehelin, 1963; Shore, Busfield & Alpers, 1964). These findings are particularly important, since they may explain some o the pharmacological properties of the α-methylated amino acids, which were originally thought to act as decarboxylase inhibitors (Sourkes, 1954). Muscholl & Maître (1963) have shown that α-methyl noradrenaline can be released from the rabbit heart by stimulation of the sympathetic nerve supply. This leads to the important concept that it may be possible to replace the normal transmitter substance with other structurally related 'false neurotransmitters'. This concept will be discussed in more detail in chapter 6. For a review of the metabolism of α-methyl amino acids and their derivatives see Muscholl (1966).

4. The cellular localization of catecholamine biosynthesis

The enzymes involved in the biosynthesis of noradrenaline have been shown to be present in peripheral tissues, in sympathetic ganglia and axons, in the adrenal medulla and in the brain. In all these tissues the enzymes are thought to be present exclusively in those cells which normally contain catecholamines (adrenal medullary chromaffin cells, postganglionic sympathetic neurones and catecholamine-containing neurones in the brain). The aromatic L-amino acid decarboxylase activity of certain peripheral tissues is greatly reduced after section and degeneration of the sympathetic nerve supply to these tissues (Andén, Magnusson & Rosengren, 1965) or in the tissues of immunosympathectomized animals (Klingman, 1965; Iversen, Glowinski & Axelrod, 1966). The ability of peripheral tissues to β-hydroxylate labelled tyramine or α-methyl tyramine is also reduced by sympathetic denervation (Carlsson & Waldeck, 1963*a*; Fischer, Musacchio, Kopin & Axelrod, 1964) or immunosympathectomy (Iversen *et al.* 1966), suggesting a similar association of dopamine-β-hydroxylase with sympathetic nerves. In brain homogenates aromatic L-amino acid decarboxylase and tyrosine hydroxylase have been shown to be associated with fractions containing pinched-off nerve terminals (synaptosomes) (De Robertis, 1964; McGeer, Bagchi & McGeer, 1965). In homogenates of peripheral tissues tyrosine hydroxylase and dopamine-β-hydroxylase are associated with particulate fractions. Dopamine-β-hydroxylase appears to be present in noradrenaline storage particles released from adrenergic terminals during the homogenization of such peripheral tissues. Aromatic L-amino acid decarboxylase activity is largely found in the supernatant fraction after centrifugation of tissue homogenates. However, recent evidence (Stjärne, 1966) suggests that this enzyme may also be loosely associated with intraneuronal storage particles in sympathetic terminals. Purified preparations of such particles from splenic nerves are able to synthesize noradrenaline from labelled DOPA, but are not able to utilize tyrosine as a substrate (Stjärne, 1966), suggesting that the enzyme catalysing the first step of the synthetic pathway, tyrosine hydroxylase, is not present in the storage particles. The nature of the particulate fraction with which tyrosine hydroxylase is associated is obscure, but it may possibly represent fragments of the axonal membrane or microsomal elements.

The site of synthesis of the noradrenaline present in the terminals of sympathetic nerves is not known. The presence of the synthetic enzymes in sympathetic ganglia and axons, together with recent evidence for a rapid proximo-distal flow in such axons has suggested the possibility that noradrenaline is synthesized in the perikarya of sympathetic neurones and is transported in storage particles down the sympathetic axons to the terminals (Carlsson, 1966; Dahlström, 1965). However, previous studies with isolated organs have shown that the terminal regions of sympathetic nerves, when removed from the body and hence devoid of all connection with their cell bodies, are able to sustain rapid rates of noradrenaline synthesis when perfused with suitable precursors (Levitt, Spector, Sjoerdsma & Udenfriend, 1965). Since the sympathetic terminals clearly contain the necessary enzymic machinery to synthesize noradrenaline at rates comparable to those found *in vivo*, it seems unnecessary to suppose that the synthesis of noradrenaline in the cell bodies of these neurones has much importance for supplying the transmitter to the terminals. The existence of a proximo–distal axonal flow of materials from the cell body to the nerve terminals is a general property of neurones (Weiss, 1961). The function of such a proximo–distal flow may be to replace the supply of enzymes and macromolecular structural components at the nerve terminals, rather than to supply the transmitter itself in a ready made form.

5. Rate of turnover of noradrenaline *in vivo*

Estimates of the rate of synthesis of noradrenaline in sympathetic nerves have been made with a variety of experimental techniques. Perhaps the most satisfactory approach has been to introduce labelled noradrenaline into the endogenous amine stores in peripheral tissues or brain by the administration of labelled tyrosine or DOPA and then to observe the rate at which the radioactive noradrenaline becomes diluted with newly synthesized endogenous catecholamine (Udenfriend & Zaltzman-Nirenberg, 1963; Burack & Draskóczy, 1964) (Fig. 5.2). An alternative approach is to label the tissue stores of noradrenaline by the administration of small doses of radioactive nor-

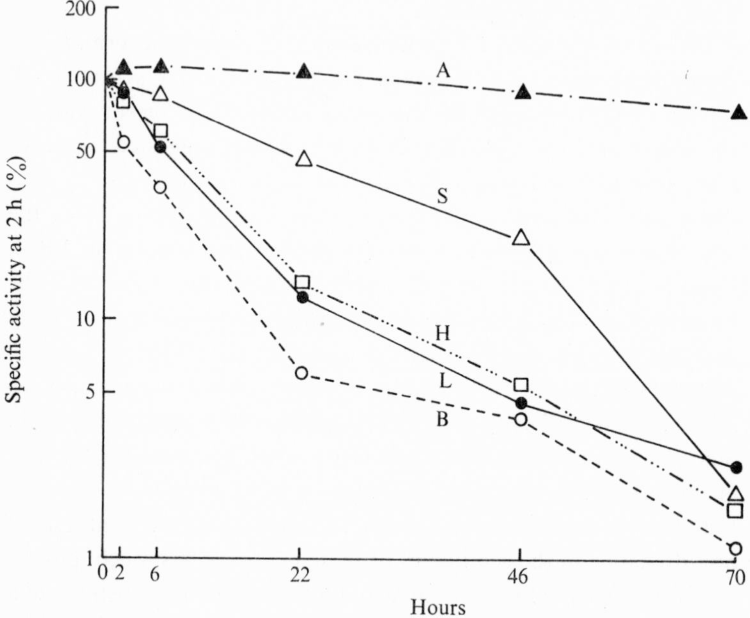

Fig. 5.2. Rate of turnover of catecholamines *in vivo*. Comparison of the rate of decline in specific activity of catecholamines in the adrenal glands (A), spleen (S), heart (H), liver (L) and brain (B) of mice. Zero time is 2 h after the intravenous injection of H^3-DOPA. After this time H^3-noradrenaline is the major radioactive catecholamine in all tissues, except the spleen, so that the curves effectively represent the rate of turnover of noradrenaline. This is true also for the spleen after 8 h. Values are the means for 3–8 determinations. (From Burack & Draskóczy, 1964.)

adrenaline and then to follow the rate of disappearance of the radioactive material (Montanari, Costa, Beaven & Brodie, 1963). If radioactive noradrenaline is injected directly into the lateral ventricle of the brain, this method can also be used to measure the rate of turnover of noradrenaline in the central nervous system (Iversen & Glowinski, 1966). In peripheral sympathetic nerves and brain the rate of noradrenaline synthesis amounts to 5–15 % of the total tissue content per hour (Montanari *et al.* 1963; Udenfriend & Zaltzman-Nirenberg, 1963). The availability of effective inhibitors of noradrenaline biosynthesis, such as α-methyl-*p*-tyrosine, offers another method for measuring noradrenaline synthesis. Since the rate of synthesis normally is balanced by the rate of amine utilization, turnover rates can be measured after the blockade of biosynthesis simply by estimating the rate of disappearance of noradrenaline from tissue stores (Spector, Sjoerdsma & Udenfriend, 1965; Iversen & Glowinski, 1966).

Rates of catecholamine turnover have also been estimated by less direct methods. Spector, Melmon & Sjoerdsma (1962) used the rate of recovery of the noradrenaline content of the heart and brain of rats after treatment with a short-acting depleting agent, and estimated the rate of noradrenaline turnover in these tissues to be 10–15 % of the total content per hour. Green & Erickson (1960) measured the rate of noradrenaline accumulation in the rat brain after inhibition of monoamine oxidase with tranylcypromine and estimated a turnover of 17 % of the total content per hour. A similar method was used by Ehringer & Hornkiewicz (1960) who reported a rapid turnover of dopamine in the striatum.

Little is known of the physiological factors controlling the rate of noradrenaline synthesis. It is possible that very high rates of noradrenaline synthesis may occur under certain conditions. No signs of fatigue can be detected after the prolonged stimulation of sympathetic nerves at high frequencies (Orias, 1932; Dye, 1935; Luco & Goni, 1948). It is possible that a rapid synthesis of transmitter occurs under such conditions. However, the unique phenomenon of transmitter re-use in adrenergic nerves may be a more important factor in the resistance of adrenergic synapses to fatigue.

6. Rate-limiting step in noradrenaline biosynthesis

Knowledge of the rate-limiting step in the tyrosine to noradrenaline pathway is essential for an understanding of the regulation of this sequence of reactions. It is at this point that the most effective and sensitive physiological and pharmacological controls may be exerted. Recent evidence suggests that the pacemaker for noradrenaline synthesis is the first step in the pathway—the conversion of tyrosine to DOPA catalysed by tyrosine hydroxylase. This evidence has been recently reviewed (Levitt *et al.* 1965; Kaufman & Friedman, 1965) and briefly is as follows. In isolated guinea-pig hearts perfused with DOPA or dopamine, high rates of noradrenaline synthesis can be attained (as much as 75 % of the total noradrenaline content per hour)

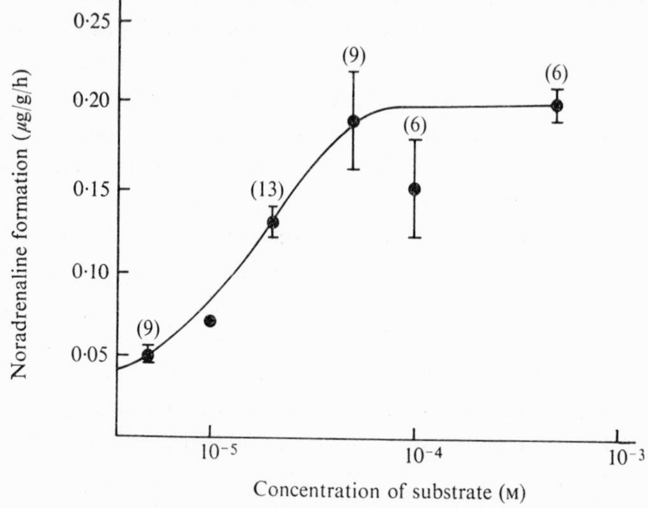

Fig. 5.3. Rate-limiting step in noradrenaline biosynthesis. As the concentration of radioactive L-tyrosine in the perfusing fluid is raised the rate of formation of radioactive noradrenaline in the isolated guinea-pig heart increases to reach a saturation level. In contrast, no such saturation in the rate of noradrenaline formation was seen with increasing concentrations of DOPA or dopamine in the perfusing fluid. Each point is the mean for 6–13 experiments (numbers in parentheses), vertical lines indicate standard errors of the mean. (From Levitt, Spector, Sjoerdsma & Udenfriend, 1965.)

and the rate of noradrenaline synthesis increases linearly with increasing concentrations of DOPA or dopamine. With tyrosine as a substrate, however, the maximum rate of noradrenaline synthesis is much lower (about 10 % of the noradrenaline content per hour) and this rate does not further increase at high concentrations of tyrosine (Fig. 5.3). The K_m for the conversion of tyrosine to noradrenaline in such a system is the same as that found for the conversion of tyrosine to DOPA by tyrosine hydroxylase. Furthermore the activity of tyrosine hydroxylase is always much lower than that of the other enzymes in the synthetic pathway. In the adrenal medulla, for example, the activity of tyrosine hydroxylase is more than 1 000 times lower than that of aromatic L-amino acid decarboxylase or dopamine-β-hydroxylase. The maximum activity of tyrosine hydroxylase in this tissue corresponds closely to the rate of noradrenaline and adrenaline synthesis observed *in vivo*. The concentration of free tyrosine in the adrenal medulla and peripheral tissues is of the order of 10^{-4} M, whilst the concentrations of DOPA and dopamine in adrenergic nerves are generally too low to be measurable. There is thus considerable evidence to suggest that tyrosine hydroxylase is the rate-limiting step for noradrenaline synthesis.

The inhibition of tyrosine hydroxylase activity by high concentrations of the final product of the synthetic pathway, L-noradrenaline, may furthermore represent a mechanism for the feedback control of noradrenaline synthesis *in vivo* (Udenfriend, 1966; see also Wurtman & Axelrod, 1966).

7. The use of synthesis inhibitors

The availability of effective inhibitors of noradrenaline biosynthesis would be of great importance both for use as tools in research and possibly as clinically useful drugs. Hitherto, much interest has centred on the possibility of inhibiting the decarboxylation of DOPA. It was at first thought that the fall in the levels of endogenous noradrenaline in brain and peripheral tissues after treatment with α-methyl-DOPA and α-methyl-*m*-tyrosine was caused by an inhibition of decarboxylation. But it is now believed that the inhibition of noradrenaline synthesis plays little role in the amine-depleting actions of these compounds, as discussed above. For example, several compounds which are more potent than the α-methyl amino-acids as *in vitro* inhibitors of DOPA decarboxylation fail to produce a fall in the noradrenaline content of tissue when administered *in vivo* (Drain, Horlington, Lazare & Poulter, 1962). Because of the broad specificity of aromatic-L-amino acid decarboxylase and its relative abundance in most tissues, this is clearly not a good point at which to inhibit catecholamine biosynthesis. Similarly there has been little success in obtaining an effective inhibitor of dopamine-β-hydroxylase, although several *in vitro* inhibitors of this enzyme have been described (Nikodijevic, Creveling & Udenfriend, 1963; Van der Schoot & Creveling, 1965). The recent discovery of tyrosine hydroxylase as the rate-limiting step in noradrenaline biosynthesis, however, has led to the development of several effective inhibitors of this enzyme. Certain iodo-tyrosine derivatives and L-α-methyl-*p*-tyrosine (Fig. 5.4) are potent inhibitors of the enzyme both *in vitro* and *in vivo* (Goldstein & Weiss, 1965; Spector, Sjoerdsma & Udenfriend, 1965; Spector, Mater, Sjoerdsma & Udenfriend, 1965; Udenfriend, Zaltzman-Nirenberg & Nagatsu, 1965). The levels of noradrenaline

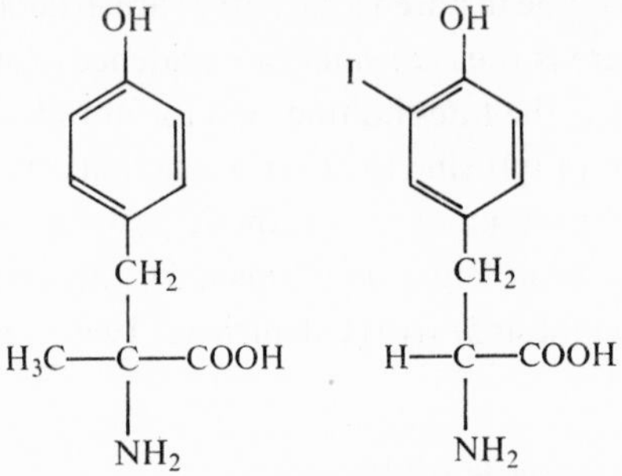

OH, CH_2, H_3C—C—COOH, NH_2

α-Methyl-*p*-tyrosine

OH, I, CH_2, H—C—COOH, NH_2

3-Iodotyrosine

Fig. 5.4. Inhibitors of noradrenaline biosynthesis. Structures of two potent inhibitors of tyrosine hydroxylase.

in peripheral tissue and brain fall after the administration of 3-iodo-L-tyrosine, 3-iodo-α-methyl-DL-tyrosine, 3,5-diiodo-L-tyrosine or L-α-methyl-*p*-tyrosine, while the levels of serotonin remain unaffected. The iodo-tyrosines are particularly interesting since both 3-iodo-tyrosine and 3,5-diiodo-tyrosine are normally produced by the thyroid gland during the synthesis of the thyroid hormones. These new drugs will certainly prove to be of great interest for research on noradrenergic mechanisms: with such drugs it may be possible to produce a well-defined and complete 'chemical sympathectomy'.

B. THE METABOLIC DEGRADATION OF CATECHOLAMINES

1. Introduction

In the last decade great advances have been made in elucidating the enzymic pathways of noradrenaline and adrenaline metabolism. The literature is now so extensive that no more than a brief survey will be given here; for more detailed reviews see Axelrod (1959), Iisalo (1962) and Kopin (1964). The present summary will be concerned mainly with the metabolic fate of noradrenaline after release from sympathetic nerves.

2. Monoamine oxidase

An enzyme in liver which oxidized tyramine was described by Hare in 1928. Blaschko, Richter & Schlossman (1937) later showed that the liver enzyme in rats, guinea-pigs and rabbits would also destroy adrenaline. This activity was also found in the kidneys of these animals. The enzyme was called 'amine oxidase' by Richter (1937), who showed it to be capable of the oxidative deamination of many amines. The name 'monoamine oxidase' has been adopted to distinguish the enzyme which catalyses the oxidative deamination of catecholamines and other monoamines from other enzymes which specifically attack diamines, such as spermine oxidase or histaminase (Hagen & Weiner, 1959).

Monoamine oxidase is distributed widely in animal tissues, being particularly abundant in liver, kidney, intestine and central nervous system. Appreciable monoamine oxidase activity is associated with all parts of the sympathetic nervous system. The enzyme is not well characterized (Blaschko, 1952) owing to the difficulty of extracting it in a soluble form from tissues, where it is exclusively localized in mitochondria (Hawkins, 1952; De Lores Arnaiz & De Robertis, 1962). The enzyme seems to be a flavoprotein which forms the aldehyde derivatives of monoamines. The subsequent oxidation of the aldehyde derivatives is catalysed either by an NAD-linked dehydrogenase or by another flavoprotein ('aldehyde oxidase'). The intermediate aldehyde derivatives formed during the metabolism of adrenaline and noradrenaline may also be reduced rather than oxidized, and then form the corresponding glycol derivatives (Fig. 5.7).

Of the physiologically important substrates for monoamine oxidase, dopamine and serotonin are attacked more rapidly than adrenaline and noradrenaline. The substrate specificity of the enzyme has been studied extensively (Blaschko *et al.* 1937; Pratesi & Blaschko, 1959; Blaschko, 1952). The enzyme has a high affinity for phenylethylamine derivatives which lack a β-hydroxyl substituent and which have monophenolic or catechol substituents in the aromatic ring. Thus dopamine, tyramine and β-phenylethylamine and many other sympathomimetic amines are good substrates for the enzyme. The enzyme in most species shows no stereochemical specificity for the optical enantiomers of adrenaline and noradrenaline (Pratesi & Blaschko,

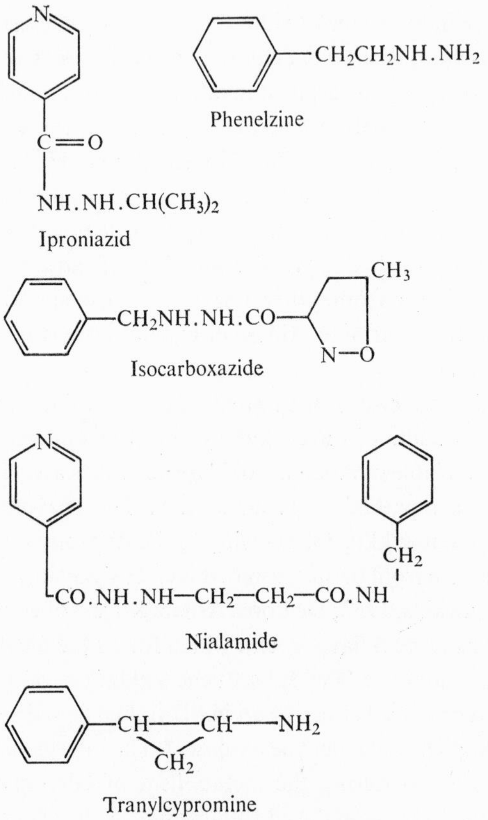

N

C=O

$NH.NH.CH(CH_3)_2$

Iproniazid

$CH_2CH_2NH.NH_2$

Phenelzine

CH_3

$CH_2NH.NH.CO$

N—O

Isocarboxazide

N

CH_2

$CO.NH.NH—CH_2—CH_2—CO.NH$

Nialamide

CH——CH——NH_2

CH_2

Tranylcypromine

Fig. 5.5. Some inhibitors of monoamine oxidase.

1959). Phenylethylamine derivatives with an α-methyl substituent are not attacked by the enzyme, but act as competitive inhibitors; e.g. amphetamine, ephedrine, methamphetamine, α-methylnoradrenaline (Blaschko *et al.* 1937; Blaschko, 1938). Adrenaline and noradrenaline are attacked at approximately equal rates.

Many potent inhibitors of monoamine oxidase have been introduced in recent years, and a voluminous literature exists describing their pharmacological and biochemical effects (Zeller, 1959; Pletscher, 1961). These compounds are now widely used for the therapy of psychic depression, angina pectoris and hypertension. Since the discovery of the monoamine oxidase inhibitory powers of iproniazid (Zeller & Barsky, 1952) many other inhibitors have been developed. These can be divided into compounds which produce a long-lasting inhibition by an irreversible non-competitive inhibition of the enzyme *in vitro* and *in vivo* (this group includes many hydrazine derivatives related to iproniazid) and those which produce a short-lasting reversible inhibition of the enzyme, such as the harmala alkaloids harmine and harmaline (Pletscher, Besendorf, Bächtold & Gey, 1959). The structures of some of these compounds are shown in Fig. 5.5.

3. Catechol-*O*-methyl transferase

An enzyme was discovered by Axelrod (1957) which *O*-methylates adrenaline, noradrenaline and other catechols. The enzyme catalyses the transfer of methyl groups from *S*-adenosyl methionine to the *meta*-hydroxyl group of such compounds. Catechol-*O*-methyl transferase is distributed widely in the cell sap of animal tissues, being particularly abundant in liver and kidney (Axelrod, Albers & Clemente, 1959; Axelrod & Tomchick, 1958). Appreciable enzyme activities were also found in the central nervous system and in many peripheral tissues such as salivary glands, heart and spleen. The purified enzyme requires *S*-adenosyl methionine and Mg^{2+} ions for activity and will *O*-methylate most catechol substrates. Pyrogallol acts as a competitive inhibitor *in vitro* and *in vivo* and more potent inhibitors have been described including certain tropolone, dopacetamide and papaverine derivatives (Belleau & Burba, 1961; Carlsson, Lindqvist, Fila-Hromadko & Corrodi, 1962; Ross & Haljasmaa, 1964*a*,*b*; Burba & Murnaghan, 1965). The structures of some of these inhibitors are shown in Fig. 5.6.

4. Metabolism of catecholamines *in vivo*

There remains considerable uncertainty about the relative importance of *O*-methylation or oxidative deamination in the metabolism of catecholamines under physiological conditions. Experiments by

Axelrod and his co-workers, in which the metabolic fate of intravenously administered radioactive catecholamines was studied, showed that *O*-methylation was the major metabolic pathway for circulating adrenaline and noradrenaline. The *O*-methylated compounds

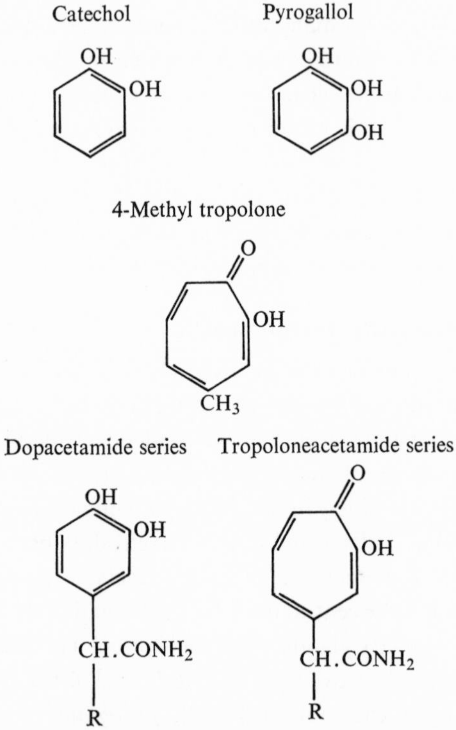
Catechol
Pyrogallol
OH
OH
OH
OH
OH
4-Methyl tropolone
O
OH
CH3
Dopacetamide series
Tropoloneacetamide series
OH
OH
O
OH
CH.CONH2
R
CH.CONH2
R

Fig. 5.6. Some inhibitors of catechol-*O*-methyl-transferase.

metanephrine and normetanephrine thus formed are subsequently deaminated by monoamine oxidase to yield the major urinary excretion product, 3-methoxy-4-hydroxymandelic acid (VMA). The pathways of metabolism open to adrenaline and noradrenaline are thus now more clearly understood, as indicated in Fig. 5.7 and Table 5.1 (Axelrod, Inscoe, Senoh & Witkop, 1958; Axelrod, 1959, 1960; Axelrod, Weil-Malherbe & Tomchick, 1959; Whitby, Axelrod & Weil-Malherbe, 1961; Kopin, Axelrod & Gordon, 1961; La Brosse, Axelrod, Kopin & Kety, 1961). The metabolism of circulating cate-

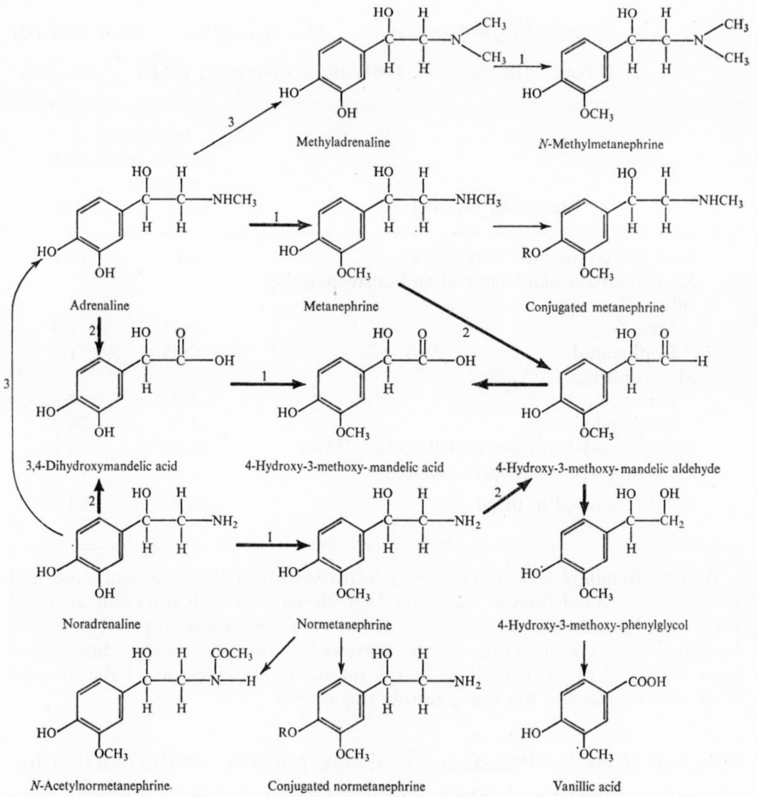

Methyladrenaline
N-Methylmetanephrine
Adrenaline
Metanephrine
Conjugated metanephrine
3,4-Dihydroxymandelic acid
4-Hydroxy-3-methoxy-mandelic acid
4-Hydroxy-3-methoxy-mandelic aldehyde
Noradrenaline
Normetanephrine
4-Hydroxy-3-methoxy-phenylglycol
N-Acetylnormetanephrine
Conjugated normetanephrine
Vanillic acid

Fig. 5.7. Pathways for the metabolism of noradrenaline and adrenaline. 1 = catechol-*O*-methyl transferase; 2 = monoamine oxidase; 3 = phenylethanolamine *N*-methyl transferase. Major pathways indicated by heavy arrows. (From Axelrod, 1965.)

cholamines takes place mainly in the liver, which is a particularly abundant source of the transferase (Axelrod, 1959; De Schraepdryver & Kirshner, 1961; Crout, Creveling & Udenfriend, 1961). Carlsson & Waldeck (1963*b*) showed that the importance of *O*-methylation for the inactivation of labelled noradrenaline in the mouse depended on the route of injection. Doses of H^3-noradrenaline administered intraperitoneally which were absorbed into the portal hepatic circulation were inactivated almost exclusively by *O*-methylation, whereas intravenous doses were inactivated largely by other mechanisms. Liver

Table 5.1. *Metabolic fate of injected* H^3*-adrenaline in man and rat*

(From LaBrosse *et al.* 1961, and Kopin *et al.* 1961.)

Metabolic product	Percentage of administered dose: Man	Rat
Free and conjugated adrenaline	6·5	14·1
3,4-Dihydroxymandelic acid and corresponding glycol		
Free	0·85	1·4
Conjugated	0·75	1·2
Metanephrine		
Free	5·0	4·8
Conjugated	33·7	28·3
3-Methoxy-4-hydroxymandelic acid (VMA)	39·2	6·2
3-Methoxy-4-hydroxyphenylglycol	6·8	17·6
Total recovered in urine	94·0	73·6

DL-H^3-adrenaline was administered intravenously to humans and rats, and urine was collected for the following 54 h (human) or 48 h (rat) and analysed for tritiated metabolites of adrenaline. Results are mean values for twelve normal human subjects and eight rats, and are expressed as percentage of the administered dose recovered in each fraction. In the rat significant amounts of the injected dose were excreted in the faeces during this period.

catechol-*O*-methyl transferase is also responsible for the inactivation of orally administered catecholamines (Resnick, 1963); this may have some importance in view of the large amounts of noradrenaline found in certain vegetable tissues. It thus seems well established that *O*-methylation is the major metabolic route for the inactivation of circulating catecholamines. Under normal conditions these would derive mainly from the adrenal medulla and to some extent from the overflow of noradrenaline released at peripheral sympathetic nerve terminals. However, the rapid inactivation of circulating catecholamines does not depend on the activity of either catechol-*O*-methyl transferase or monoamine oxidase. Even if both enzymes are inhibited simultaneously the pressor effects of intravenously administered catecholamines are not greatly potentiated or prolonged (Crout, 1961). The mechanism involved in the rapid termination of the physiological effects of circulating catecholamines involves the transfer of the active amines to inert storage sites in peripheral tissues without involving any metabolic degradation of the catecholamines (see chapter 7).

The role of *O*-methylation in the metabolism of noradrenaline after its release from sympathetic nerve endings in peripheral tissues is far less clear. Studies of the metabolism of noradrenaline in brain and heart homogenates *in vitro* suggested that *O*-methylation was of only minor significance when compared to the active monoamine oxidase present in these tissues (Crout *et al.* 1961). It is doubtful, however, whether this approach can give a true picture of the metabolism of catecholamines under physiological conditions, since the barriers which undoubtedly exist in the intact tissue between the enzymes and their substrates are destroyed during homogenization. Studies of the fate of radioactively labelled noradrenaline in intact tissues have indicated that *O*-methylation is an important pathway for the metabolism of noradrenaline in such tissues. H^3-noradrenaline released from sympathetic nerve terminals by nerve stimulation, by displacing agents such as tyramine or by ganglionic stimulating substances such as dimethylphenylpiperazinium iodide (DMPP) is largely metabolized by *O*-methylation (Hertting & Axelrod, 1961; Chidsey, Kahler, Kelminson & Braunwald, 1963; Kopin & Gordon, 1962, 1963; Rosell, Kopin & Axelrod, 1963).

The *O*-methylated products metanephrine and normetanephrine have less than 1 % of the pressor activity of the parent catecholamines; although the *O*-methylated amines have been reported to sensitize tissues to the actions of noradrenaline and adrenaline (Champagne, D'Iorio & Beaulnes, 1960; Bacq & Renson, 1961). After the inhibition of catechol-*O*-methyl transferase by pyrogallol the effects of adrenaline on isolated smooth muscle preparations were found to be potentiated (Bacq, Gosselin, Dresse & Renson, 1959). Pyrogallol also prolonged the pressor effects of injected adrenaline in the cat and increased the toxicity of adrenaline injections in the rat; these effects were much less pronounced for noradrenaline (Wylie, Archer & Arnold, 1960). Similar effects have been found with 4-methyl tropolone (Murnaghan & Mazurkiewicz, 1963). The inhibition of catechol-*O*-methyl transferase, while having little effect on the actions of noradrenaline, does potentiate and prolong the actions of injected adrenaline and this effect is even more marked with *N*-ethyladrenaline or isoprenaline (Izquierdo & Kaumann, 1963; Ross, 1963). The evidence suggests that *O*-methylation is an important route for the metabolism of catecholamines in blood and in the extracellular space of tissues, but there is little evidence to suggest that *O*-methylation plays any role in the intraneuronal metabolism of noradrenaline in sympathetic nerves. The activity of catechol-*O*-methyl transferase in peripheral tissues is not affected by surgical sympathectomy and is unchanged in the tissues of immunosympathectomized rats (Waltman & Sears, 1964; Potter, Cooper, Willman & Wolfe, 1965; Iversen *et al.* 1966). The enzyme thus does not seem to be present in sympathetic nerves, but is presumably located in some adjacent cells, possibly in the post-synaptic effector cells, in the region of adrenergic synapses. Catechol-*O*-methyl transferase activity is associated with synaptosomes isolated from rat brain homogenates (Alberici, De Lores Arnaiz & De Robertis, 1965); since postsynaptic membrane fragments are integral parts of the synaptosomal particles this does not rule out a post-synaptic localization of the enzyme in the brain as in the peripheral nervous system. *O*-Methylation is also important in the metabolism of noradrenaline in the brain. Mannarino, Kirshner & Nashold (1963) found that *O*-methylated compounds were the principle metabolites of C^{14}-noradrenaline injected into the lateral ventricle of the cat brain. After the injection of small amounts of H^3-noradrenaline into the lateral ventricle of the rat brain Glowinski, Kopin & Axelrod (1965) also found that *O*-methylation played an important role in the metabolism of the injected amine. Four hours after the injection of H^3-noradrenaline analysis of the remaining radioactivity in the brain showed that unchanged H^3-noradrenaline accounted for approximately 50 %, H^3-normetanephrine for 14 %, deaminated catechol metabolites for 3 %, and *O*-methylated deaminated metabolites for 30 %. The major *O*-methylated deaminated metabolite in the brain was 3-methoxy-4-hydroxyphenylglycol rather than VMA.

Monoamine oxidase was considered for a long time to be the enzyme chiefly concerned with the metabolic inactivation of catecholamines. It was suggested that this enzyme played a role at sympathetic nerve endings analogous to that of acetylcholinesterase at cholinergic nerve endings (Burn, 1952). This assumption rested largely on the finding that the monoamine oxidase activity of certain peripheral tissues appeared to decrease after sympathetic denervation of these tissues (Burn & Robinson, 1952). Subsequent investigations, however, failed to confirm this finding (Armin, Grant, Thompson & Tickner, 1953; Burn, Philpot & Trendelenburg, 1954) and showed that monoamine oxidase activity persisted undiminished in sympathetically denervated tissues. More recent findings have demonstrated a fall in monoamine oxidase activity after sympathetic denervation of the rat pineal gland or the rabbit iris (Snyder, Fischer & Axelrod, 1965; Waltman & Sears, 1964), suggesting that at least part of the monoamine oxidase activity in peripheral tissues is associated with sympathetic nerves. In most tissues, however, the amount of monoamine oxidase in sympathetic nerves is so small in comparison to the amounts in other cells that little or no change is observed on denervation or in immunosympathectomized animals (Iversen *et al.* 1966). In contrast to the specific localization of acetylcholinesterase in the cholinergic synapse, there seems to be no such selective association of monoamine oxidase with adrenergic synapses (Koelle & Valk, 1954).

Studies of the fate of injected H^3-noradrenaline and H^3-adrenaline (quoted above) showed that monamine oxidase played only a secondary role to catechol-*O*-methyl transferase in the metabolic degradation of circulating catecholamines. Oxidative deamination also seemed to be of little importance in the metabolism of labelled noradrenaline released from sympathetic nerve endings. The inhibition of monoamine oxidase with iproniazid did not prolong or potentiate the actions of catecholamines on the cat nictitating membrane (Griesemer *et al.* 1953), and this important finding has since been repeated by many others with different monoamine oxidase inhibitors and other tissues. Monoamine oxidase inhibition failed to potentiate the actions of catecholamines administered exogenously or released by nerve stimulation (Balzer & Holtz, 1956; Corne & Graham, 1957; Kamijo, Koelle & Wagner, 1955). Monoamine oxidase inhibitors also failed to affect the rate of excretion or inactivation of administered catecholamines (Celander & Mellander, 1955; Friend, Zileli, Hamilin & Reutter, 1958). Monoamine oxidase inhibition also failed to increase the outflow of noradrenaline from the cat spleen during stimulation of the sympathetic nerve supply (Brown & Gillespie, 1957). There is thus little reason to believe that monoamine oxidase plays any important role in the inactivation of noradrenaline in peripheral tissues (Axelrod, 1959). The activity of this enzyme does markedly affect the actions of certain other sympathomimetic amines in tissues; thus the inhibition of monoamine oxidase produces a considerable potentiation of the pharmacological actions of β-phenylethylamine and tyramine, which are normally rapidly destroyed by deamination (Griesemer *et al.* 1953).

There is, however, considerable evidence that monoamine oxidase plays some role in the intraneuronal metabolism of noradrenaline (Kopin, 1964). The endogenous levels of noradrenaline and dopamine in the brain are greatly increased in many species after the administration of monoamine oxidase inhibitors such as iproniazid (Spector, Maling & Shore, 1959; Spector, Shore & Brodie, 1960; Pletscher, 1961) and similar rises in the endogenous noradrenaline content of peripheral tissues are also found (Pletscher, 1961; Goldberg & Shideman, 1962; Bhagat, 1963). During treatment with an inhibitor of monoamine oxidase the endogenous noradrenaline content of the brain and peripheral tissues may rise two or three times above the normal levels. It has been suggested that intraneuronal monoamine oxidase may serve to regulate the storage levels of noradrenaline in sympathetic nerve endings by destroying noradrenaline synthesized in excess of the normal storage capacity (Brodie, Spector & Shore, 1959; Brodie & Beaven, 1963). Further evidence in favour of the view that monoamine oxidase plays an important role in the neuronal metabolism of noradrenaline was provided by Kopin & Gordon (1962, 1963). They found that after the intravenous administration of H^3-noradrenaline the labelled amine disappeared from rat tissues in a multiphasic fashion. The first phase of rapid release during the first 4 h after injection was of material released from the tissues as free noradrenaline or as normetanephrine. During the second slower phase of release, from 4 to 24 h after injection, the labelled amine was released from the tissues largely in the form of deaminated metabolites (Table 5.2). This was taken to suggest the existence of two storage forms of noradrenaline in peripheral tissues: one of which was released spontaneously or by tyramine and which was largely metabolized by *O*-methylation, and a second more stable form which was normally released only slowly, which was largely metabolized by

Table 5.2. *Metabolic fate of* H³-*noradrenaline in the rat*

(From Kopin & Gordon, 1963.)

	0–3 h after administration	10–13 h after administration
Total excreted	70·2*	3·94*
Norepinephrine		
Free	23·2	1·2
Conjugated	3·1	5·2
Deaminated catechols		
Free	0·7	2·9
Conjugated	1·2	4·1
Normetanephrine		
Free	12·1	3·1
Conjugated	17·1	8·4
3-Methoxy-4-hydroxymandelic acid (VMA)	4·0	13·9
3-Methoxy-4-hydroxyphenylglycol	32·9	47·6
Total urinary tritium recovered in metabolite fractions	94·3	86·4

* Percentage of administered dose.

Rats received 100 μc DL-H³-noradrenaline intravenously and urine was collected 0–3 h after and 10–13 h after and analysed for tritiated metabolites. Results are mean values for seven experiments and are expressed as percentage of total radioactivity excreted during the collection period.

monoamine oxidase. After the inhibition of monoamine oxidase the second slower phase of H³-noradrenaline release was greatly retarded (Axelrod, Hertting & Patrick, 1961).

In certain species the inhibition of monoamine oxidase fails to produce the large rises in the endogenous noradrenaline content of tissues noted above. For example, in cats pheniprazine failed to produce a rise in the brain content of noradrenaline, although the same compound produced a marked rise in the brain levels of noradrenaline in the rabbit. Conversely, pheniprazine caused an elevation in the noradrenaline content of sympathetic ganglia in cats, although this effect was much less marked in the rabbit (Sanan & Vogt, 1962). Inhibitors of monoamine oxidase also fail to produce a rise in the heart levels of noradrenaline in dogs or cats (Maling, Highman & Spector, 1962) or in mice (Le Roy & De Schaepdryver, 1961; Sharman, Vanov & Vogt, 1962).

Monoamine oxidase does not appear to be specifically associated with adrenergic nerves, the enzyme plays a major role in the metabolism of other amines in the body, notably in serotonin metabolism. It seems probable that monoamine oxidase plays only a relatively non-specific role in controlling the noradrenaline content of peripheral catecholamine stores.

In summary, *O*-methylation is the major metabolic pathway for extracellular catecholamines; oxidative deamination may be connected in some poorly understood manner with the regulation of the noradrenaline content of tissue stores. Neither metabolic route is essential for the termination of the physiological actions of catecholamines *in vivo*, as shown by the failure of inhibitors of either enzyme to produce a marked potentiation of the actions of noradrenaline or adrenaline after their administration *in vivo* (Crout, 1961) or by the failure of metabolic inhibitors to increase the overflow of neuronally released noradrenaline in stimulated tissues (Stinson, 1961; Brown & Gillespie, 1957). The inactivation of noradrenaline, unlike the inactivation of acetylcholine at cholinergic synapses, does not appear to involve the metabolic degradation of the transmitter substance.

REFERENCES

Alberici, M., De Lores Arnaiz, G. R. & De Robertis, E. (1965). Catechol-*O*-methyl transferase in nerve endings of rat brain. *Life Sciences*, **4**, 1951–1960.

Andén, N. E., Magnusson, T. & Rosengren, E. (1965). Occurrence of dihydroxyphenylalanine decarboxylase in nerves of the spinal cord and sympathetically innervated organs. *Acta physiol. scand.* **64**, 127–135.

Armin, J., Grant, R. T., Thompson, R. H. S. & Tickner, A. (1953). An explanation for the heightened vascular reactivity of the denervated rabbit's ear. *J. Physiol., Lond.* **121**, 603–622.

Axelrod, J. (1957). *O*-Methylation of epinephrine and other catechols *in vitro* and *in vivo*. *Science* **126**, 400–401.

Axelrod, J. (1959). Metabolism of norepinephrine and other sympathomimetic amines. *Physiol. Rev.* **39**, 751–776.

Axelrod, J. (1960). In *Adrenergic Mechanisms*. CIBA. London: Churchill.

Axelrod, J. (1962). Purification and properties of phenylethanolamine-*N*-methyl transferase. *J. biol. Chem.* **237**, 1557–1660.

Axelrod, J. (1965). The metabolism, storage and release of catecholamines. *Recent. Prog. Horm. Res.* **21**, 597–622.

Axelrod, J. (1966). Methylation reactions in the formation and metabolism of catecholamines and other biogenic amines. (Second International Catecholamine Symposium, Milan.) *Pharmac. Rev.* **18**, 95–113.

Axelrod, J., Albers, R. W. & Clemente, C. D. (1959). Distribution of catechol-*O*-methyl transferase in the nervous system and other tissues. *J. Neurochem.* **5**, 68–72.

Axelrod, J., Hertting, G. & Patrick, R. W. (1961). Inhibition of H^3-norepinephrine release by monoamine oxidase inhibitors. *J. Pharmac. exp. Ther.* **134**, 325–328.

Axelrod, J., Inscoe, J. K., Senoh, S. & Witkop, B. (1958). *O*-Methylation, the principal pathway for the metabolism of epinephrine and norepinephrine in the rat. *Biochim. biophys. Acta* **27**, 210–211.

Axelrod, J. & Tomchick, R. (1958). Enzymatic *O*-methylation of norepinephrine and other catechols. *J. biol. Chem.* **233**, 702–705.

Axelrod, J., Weil-Malherbe, H. & Tomchick, R. (1959). The physiological disposition of H^3-norepinephrine and its metabolite metanephrine. *J. Pharmac. exp. Ther.* **127**, 251–256.

Bacq, Z. M., Gosselin, L., Dresse, A. & Renson, J. (1959). Inhibition of *O*-methyltransferase by catechol and sensitization to epinephrine. *Science* **130**, 453–454.

Bacq, Z. M. & Renson, J. (1961). Actions et importance physiologique de la métanéphrine et de la normétanéphrine. *Archs int. Pharmacodyn Thér.* **130**, 385–402.

Bagchi, S. P. & McGeer, P. L. (1964). Some properties of tyrosine hydroxylase from the caudate nucleus. *Life Sciences* **3**, 1195–1200.

Balzer, H. & Holtz, P. (1956). Beeinflussung der Wirkung biogener Amine durch Hemmung der Aminoxydase. *Arch. exp. Path. Pharmak.* **227**, 547–558.

Belleau, B. & Burba, J. (1961). Tropolones: a unique class of potent noncompetitive inhibitors of *s*-adenosyl-*l*-methionine-catechol-methyltransferase. *Biochim. biophys. Acta* **54**, 195–196.

Bhagat, B. (1963). Effect of various monoamine oxidase inhibitors on the catecholamine content of rat heart. *Archs. int. pharmacodyn. Thér.* **146**, 65–72.

Blaschko, H. (1938). Amine oxidase and ephedrine. *J. Physiol., Lond.* **93**, 7–8*P*.

Blaschko, H. (1939). The specific action of L-DOPA decarboxylase. *J. Physiol., Lond.* **96**, 50–51*P*.

Blaschko, H. (1952). Amine oxidase and amine metabolism. *Pharmac. Rev.* **4**, 415–458.

Blaschko, H., Burn, J. H. & Langemann, H. (1950). The formation of noradrenaline from dihydroxyphenylserine. *Br. J. Pharmac. Chemother.* **5**, 431–437.

Blaschko, H., Hagen, P. & Welch, A. D. (1955). Observations on the intracellular granules of the adrenal medulla. *J. Physiol., Lond.* **129**, 27–49.

Blaschko, H., Richter, D. & Schlossman, H. (1937). The oxidation of adrenaline and other amines. *Biochem. J.* **31**, 2187–2196.

Brodie, B. B. & Beaven, M. A. (1963). Neurochemical transducer systems. *Med. exp.* **8**, 320–351.

Brodie, B. B., Spector, S. & Shore, P. A. (1959). Interaction of drugs with norepinephrine in the brain. *Pharmac. Rev.* **11**, 548–564.

Brown, G. L. & Gillespie, J. S. (1957). The output of sympathetic transmitter from the spleen of the cat. *J. Physiol., Lond.* **138**, 81–102.

Burack, W. R. & Draskóczy, P. R. (1964). The turnover of endogenously labelled catecholamines in several regions of the sympathetic nervous system. *J. Pharmac. exp. Ther.* **144**, 66–75.

Burba, J. V. & Murnaghan, M. F. (1965). Catechol-*O*-methyl transferase inhibition and potentiation of epinephrine responses by desmethylpapaverine. *Biochem. Pharmacol.* **14**, 823–829.

Burn, J. H. (1952). The enzyme at sympathetic nerve endings. *Br. med. J.* **1**, 784–787.

Burn, J. H., Philpot, F. J. & Trendelenburg, U. (1954). Effect of denervation on enzymes in iris and blood vessels. *Br. J. Pharmac. Chemother.* **9**, 423–428.

Burn, J. H. & Robinson, J. (1952). Effect of denervation on amine oxidase in structures innervated by the sympathetic. *Br. J. Pharmac. Chemother.* **7**, 304–318.

Carlsson, A. (1964). Functional significance of drug-induced changes in brain monoamine levels. *Prog. Brain Res.* **8**, 9–27.

Carlsson, A. (1966). Pharmacological depletion of catecholamine stores. (Second International Catecholamine Symposium, Milan.) *Pharmac. Rev.* **18**, 541–550.

Carlsson, A. & Lindqvist, M. (1962). *In vivo* decarboxylation of α-methyl dopa and α-methyl-*m*-tyrosine. *Acta physiol. scand.* **54**, 87–94.

Carlsson, A., Lindqvist, M., Fila-Hromadko, S. & Corrodi, H. (1962). Synthese von Catechol-*O*-Methyl Transferase-hemmenden Verbindurgen. *Helv. chim. Acta* **45**, 270–276.

Carlsson, A. & Waldeck, B. (1963*a*). β-Hydroxylation of tyramine *in vivo*. *Acta pharmac. tox.* **20**, 371–375.

Carlsson, A. & Waldeck, B. (1963*b*). On the role of the liver catechol-*O*-methyl transferase in the metabolism of circulating catecholamines. *Acta pharmac. tox.* **20**, 47–55.

Celander, O. & Mellander, S. (1955). Elimination of adrenaline and noradrenaline from the circulating blood. *Nature, Lond*, **176**, 973–974.

Champagne, J., D'Iorio, A. & Beaulnes, A. (1960). Biological activity of 3-methoxy catecholamines. *Science, N.Y.* **132**, 419–420.

Chidsey, C. A., Kahler, R. L., Kelminson, L. & Braunwald, E. (1963). Uptake and metabolism of tritiated norepinephrine in the isolated canine heart. *Circulation Res.* **12**, 220–227.

Chidsey, C. A., Kaiser, G. A. & Braunwald, E. (1963). Biosynthesis of norepinephrine in isolated canine heart. *Science, N.Y.* **139**, 828–829.

Corne, S. J. & Graham, J. D. (1957). The effect of inhibition of amine oxidase *in vivo* on administered adrenaline, noradrenaline, tyramine and serotonin. *J. Physiol., Lond.* **135**, 339–349.

Creveling, C. R., Levitt, M. & Undenfriend, S. (1962). An alternative route for the biosynthesis of norepinephrine. *Life Sciences* **1**, 523–526.

Crout, J. R. (1961). Effect of inhibiting both catechol-*O*-methyltransferase and monoamine oxidase on cardiovascular responses to norepinephrine. *Proc. Soc. exp. Biol. Med.* **108**, 482–484.

Crout, J. R., Creveling, C. R. & Udenfriend, S. (1961). Norepinephrine metabolism in rat brain and heart. *J. Pharmac. exp. Ther.* **132**, 269–277.

Dahlström, A. (1965). Observations on the accumulation of noradrenaline in the proximal and distal parts of peripheral adrenergic nerves after compression. *J. Anat.* **99**, 677–689.

De Lores Arnaiz, G. R. & De Robertis, E. (1962). Cholinergic and non-cholinergic nerve endings in the rat brain: II. *J. Neurochem.* **9**, 503–518.

Demis, D. J., Blaschko, H. & Welch, A. D. (1955). The conversion of dihydroxyphenylalanine-2-C^{14}(DOPA) to norepinephrine by bovine adrenal medullary homogenates. *J. Pharmac. exp. Ther.* **113**, 14–15.

De Robertis, E. (1964). Electron microscope and chemical study of binding sites of brain biogenic amines. *Prog. Brain Res.* **8**, 118–136.

De Schaepdryver, A. F. & Kirshner, N. (1961). The metabolism of DL-adrenaline-2-C^{14} in the cat. II. Tissue metabolism. *Archs. int. pharmacodyn. Thér.* **131**, 433–449.

Drain, D. J., Horlington, M., Lazare, R. & Poulter, G. A. (1962). The effect of α-methyl-dopa and some other decarboxylase inhibitors on brain 5-hydroxytryptamine. *Life Sciences* **1**, 93–97.

Dye, J. A. (1935). The exhaustability of the sympathin stores. *Am. J. Physiol.* **113**, 265–270.

Ehringer, H. & Hornykiewicz, O. (1960). Verteilung von Noradrenalin und Dopamin (3-hydroxytyramin) in Gehirn des Menschen und ihr Vertratten Bei Erkrankungen des Extrapyramidden Systems. *Klin. Wschr.* **38**, 1236–1239.

Fellman, J. H. (1959). Purification and properties of adrenal L-dopa decarboxylase. *Enzymologia* **20**, 366–375.

Fischer, J. E., Musacchio, J., Kopin, I. J. & Axelrod, J. (1964). Effects of denervation on the uptake and β-hydroxylation of tyramine in the rat salivary gland. *Life Sciences* **3**, 413–420.

Friedman, S. & Kaufman, S. (1965), 3,4-Dihydroxyphenylethylamine-β-hydroxylase. *J. biol. Chem.* **240**, 4763–4773.

Friend, G., Zileli, M. S., Hamilin, J. R. & Reutter, F. W. (1958). The effect of iproniazid on the inactivation of norepinephrine in the human. *J. clin. Psychopath. Psychother.* **19**, 61–68.

Gessa, G. L., Costa, E., Kuntzman, R. & Brodie, B. B. (1962). On the mechanism of norepinephrine release by α-methyl-*m*-tyrosine. *Life Sciences* **1**, 353–360.

Glowinski, J., Kopin, I. J. & Axelrod, J. (1965). Metabolism of H^3-norepinephrine in the rat brain. *J. Neurochem.* **12**, 25–30.

Goldberg, N. D. & Shideman, F. E. (1962). Species differences in the cardiac effects of a monoamine oxidase inhibitor. *J. Pharmac. exp. Ther.* **136**, 142–151.

Goldstein, M., Musacchio, J. M. & Contrera, J. F. (1962). The inhibition *in vivo* of norepinephrine synthesis by adrenalone. *Biochem. Pharmac.* **11**, 809–811.

Goldstein, M. & Weiss, Z. (1965). Inhibition of tyrosine hydroxylase by 3-iodo-L-tyrosine. *Life Sciences* **4**, 261–264.

Goodall, McC. & Kirshner, N. (1958). Biosynthesis of epinephrine and norepinephrine by sympathetic nerves and ganglia. *Circulation* **17**, 366–371.

Green, H. & Erickson, R. W. (1960). Effect of trans-2-phenylcyclopropylamine upon norepinephrine concentration and monoamine oxidase activity of rat brain. *J. Pharmac. exp. Ther.* **129**, 237–249.

Griesemer, E. C., Barsky, J., Dragstedt, C. A., Wells, J. A. & Zeller, E. A. (1953). Potentiating effects of iproniazid on the pharmacological action of sympathomimetic amines. *Proc. Soc. exp. Biol. Med.* **84**, 699–701.

Hagen, P. (1956). Biosynthesis of norepinephrine from 3,4-dihydroxyphenylethylamine (dopamine). *J. Pharmac. exp. Ther.* **116**, 26–27.

Hagen, P. (1962). Observations on the substrate specificity of dopa decarboxylase from ox adrenal medulla, human phaeochromocytoma and human argentaffinoma. *Br. J. Pharmac. Chemother.* **18**, 175–182.

Hagen, P. & Weiner, N. (1959). Enzymic oxidation of pharmacologically active amines. *Fedn. Proc.* **18**, 1005–1012.

Hare, M. L. (1928). Tyramine oxidase. *Biochem. J.* **22**, 968–979.

Hawkins, J. (1952). Localization of amine oxidase in the liver cell. *Biochem. J.* **50**, 577–581.

Hertting, G. & Axelrod, J. (1961). Fate of tritiated noradrenaline at the sympathetic nerve endings. *Nature, Lond.* **192**, 172–173.

Holtz, P. (1959). Role of L-dopa decarboxylase in the biosynthesis of catecholamines in nervous tissue and the adrenal medulla. *Pharmac. Rev.* **11**, 317–329.

Holtz, P., Heise, R. & Ludtke, K. (1938). Quantitativer Abbau von L-dioxyphenylalanin (Dopa) durch Niere. *Archs. exp. Path. Pharmak.*, **191**, 87–118.

Holtz, P. & Westermann, E. (1956) Über die Dopadecarboxylase und Histidindecarboxylase des Nervengewebes. *Arch. exp. Path. Pharmak.* **227**, 538–546.

Iisalo, E. (1962). Enzyme action on noradrenaline and adrenaline. *Acta pharmac. tox.* **19**, Suppl. 1.

Iversen, L. L. & Glowinski, J. (1966). Regional studies of catecholamines in the rat brain. II. Rate of turnover of catecholamines in various brain regions. *J. Neurochem.* **13**, 671–682.

Iversen, L. L., Glowinski, J. & Axelrod, J. (1966). The physiological disposition and metabolism of norepinephrine in immunosympathectomized animals. *J. Pharmac. exp. Ther.* **151**, 273–284.

Izquierdo, J. A. & Kaumann, A. J. (1963). Effect of pyrogallol on the duration of the cardiovascular action of catecholamines. *Archs. int. Pharmacodyn. Thér.* **144**, 437–445.

Kamijo, J., Koelle, G. B. & Wagner, H. H. (1955). Modification of the effects of sympathomimetic amines and of adrenergic nerve stimulation by 1-isonicotinyl-2-isopropylhydrazine and isonicotinic acid hydrazide. *J. Pharmac. exp. Ther.* **117**, 213–227.

Kaufman, S. & Friedman, S. (1965). Dopamine-β-hydroxylase. *Pharmac. Rev.* **17**, 71–100.

Kirshner, N. (1959). Biosynthesis of adrenaline and noradrenaline. *Pharmac. Rev.* **11**, 350–360.

Kirshner, N. & Goodall, McC. (1956). Biosynthesis of adrenaline and noradrenaline by adrenal slices. *Fedn. Proc.* **15**, 110–111.

Klingman, G. I. (1965). Catecholamine levels and dopa decarboxylase activity in peripheral organs and adrenergic tissue in the rat after immunosympathectomy. *J. Pharmac. exp. Ther.* **148**, 14–21.

Koelle, G. B. & Valk, A. de T. (1954). Physiological implications of the histochemical localization of monoamine oxidase. *J. Physiol., Lond.* **126**, 434–447.

Kopin, I. J. (1964). Storage and metabolism of catecholamines: the role of monoamine oxidase. *Pharmac. Rev.* **16**, 179–191.

Kopin, I. J., Axelrod, J. & Gordon, E. K. (1961). The metabolic fate of H^3-epinephrine and C^{14}-metanephrine in the rat. *J. biol. Chem.* **236**, 2109–2113.

Kopin, I. J. & Gordon, E. K. (1962). Metabolism of norepinephrine-H^3 released by tyramine and reserpine. *J. Pharmac. exp. Ther.* **138**, 351–359.

Kopin, I. J. & Gordon, E. K. (1963). Metabolism of administered and drug released norepinephrine-7-H^3 in the rat. *J. Pharmac. exp. Ther.* **140**, 207–216.

La Brosse, E. H., Axelrod, J., Kopin, I. J. & Kety, S. S. (1961). Metabolism of 7-H^3-epinephrine-d-bitartrate in normal young men. *J. clin. Invest.* **40**, 253–260.

Le Roy, J. G. & De Schaepdryver, A. F. (1961). Catecholamine levels of brain and heart in mice after iproniazid, syrosingopine and 10-methoxydeserpidine. *Archs. int. Pharmacodyn. Thér.* **130**, 231–234.

Levin, E. Y. & Kaufman, S. (1961). Studies on the enzyme catalysing the conversion of 3,4-dihydroxyphenylethylamine to norepinephrine. *J. biol. Chem.* **236**, 2043–2049.

Levitt, M., Spector, S., Sjoerdsma, A. & Udenfriend, S. (1965). Elucidation of the rate-limiting step in norepinephrine biosynthesis in the perfused guinea-pig heart. *J. Pharmac. exp. Ther.* **148**, 1–8.

Lovenberg, W., Weissbach, H. & Udenfriend, S. (1962). Aromatic L-amino acid decarboxylase. *J. biol. Chem.* **237**, 89–93.

Luco, J. V. & Goni, F. (1948). Synaptic fatigue and chemical mediators of post-ganglionic fibres. *J. Neurophysiol.* **11**, 497–500.

Maître, L. & Staehelin, M. (1963). Effect of α-methyl dopa on myocardial catecholamines. *Experientia* **19**, 573–575.

Maling, H. M., Highman, B. & Spector, S. (1962). Neurologic, neuropathologic and neurochemical effects of prolonged administration of phenylisopropylhydrazine (JB 516), phenylisobutylhydrazine (JB 835) and other monoamine oxidase inhibitors. *J. Pharmac. exp. Ther.* **137**, 334–343.

Mannarino, E., Kirshner, N. & Nashold, B. S. Jr. (1963). The metabolism of C^{14}-noradrenaline by cat brain. *J. Neurochem.* **10**, 373–379.

Masuoka, D. T., Clark, W. G. & Schott, H. F. (1961). Biosynthesis of catecholamines by rat brain tissue *in vitro*. *Rev. canad. Biol.* **20**, 1–6.

Masuoka, D. T., Schott, H. F., Akawie, R. I. & Clark, W. G. (1956). Conversion of C^{14}-arterenol to epinephrine *in vivo*. *Proc. Soc. exp. Biol. Med.* **93**, 5–7.

Masuoka, D. T., Schott, H. F. & Petriello, L. (1963). Formation of catecholamines by various areas of cat brain. *J. Pharmac. exp. Ther.* **139**, 73–76.

McGeer, P. L., Bagchi, S. P. & McGeer, E. G. (1965). Subcellular localization of tyrosine hydroxylase in beef caudate nucleus. *Life Sciences* **4**, 1859–1867.

McGeer, P. L. & McGeer, E. G. (1964). Formation of adrenaline by brain tissue. *Biochem. biophys. Res. Commun.* **17**, 502–507.

Milhaud, G. & Glowinski, J. (1962). Métabolisme de la dopamine-^{14}C dans le cerveau du rat. Étude du mode d'administration. *C. r. hebd. Séanc. Acad. Sci., Paris*, **255**, 203.

Montanari, R., Costa, E., Beaven, M. A. & Brodie, B. B. (1963). Turnover rates of norepinephrine in hearts of intact mice, rats and guinea-pigs using tritiated norepinephrine. *Life Sciences* **2**, 232–240.

Murnaghan, M. F. & Mazurkiewicz, I. M. (1963). Some pharmacological properties of 4-methyltropolone. *Rev. canad. Biol.* **22**, 99–102.

Musacchio, J. M. & Goldstein, M. (1963). Biosynthesis of norepinephrine and norsynephrine in the perfused rabbit heart. *Biochem. Pharmac.* **12**, 1061–1063.

Muscholl, E. (1966). Autonomic nervous system: Newer mechanisms of adrenergic blockade. *A. Rev. Pharmac.* **6**, 107–128.

Muscholl, E. & Maître, L. (1963). Release by sympathetic stimulation of α-methyl noradrenaline stored in the heart after administration of α-methyl dopa. *Experientia* **19**, 658–659.

Nagatsu, T., Levitt, M. & Udenfriend, S. (1964). The initial step in norepinephrine biosynthesis. *J. biol. Chem.* **239**, 2910–2917.

Nikodijevic, B., Creveling, C. R. & Udenfriend, S. (1963). Inhibition of dopamine-β-oxidase *in vivo* by benzyloxyamine and benzylhydrazine analogs. *J. Pharmac. exp. Ther.* **140**, 224–228.

Orias, O. (1932). Response of the nictitating membrane to prolonged stimulation of the cervical sympathetic. *Am. J. Physiol.* **102**, 87–93.

Pellerin, J. & D'Iorio, A. (1957). Metabolism of DL-3,4-dihydroxyphenylalanine-α-C^{14} in bovine adrenal homogenate. *Canad. J. Biochem. Physiol.* **35**, 151–156.

Pletscher, A. (1961). Monoaminoxydase-Hemmer. *Dt. med. Wschr.* **86**, 647–657.

Pletscher, A., Besendorf, H., Bächtold, H. P. & Gey, K. F. (1959). Über pharmakologische Beeinflussung des Zentralnervensystems durch Kurzwirkende Monoaminoxydasehemmer aus der Gruppe der Harmala-Alkaloide. *Helv. physiol. pharmac. Acta* **17**, 202–214.

Potter, L. T. & Axelrod, J. (1963). Properties of norepinephrine storage particles of the rat heart. *J. Pharmac. exp. Ther.* **142**, 299–305.

Potter, L. T., Cooper, T., Willman, V. L. & Wolfe, D. E. (1965). Synthesis, binding, release and metabolism of norepinephrine in normal and transplanted dog hearts. *Circulation Res.* **16**, 468–481.

Pratesi, P. & Blaschko, H. (1959). Specificity of amine oxidase for optically active substrates and inhibitors. *Br. J. Pharmac. Chemother.* **14**, 256–260.

Resnick, O. (1963). The metabolism of orally ingested epinephrine in man. *Life Sciences* **2**, 629–636.

Richter, D. (1937). Adrenaline and amine oxidase. *Biochem. J.* **31**, 2022–2028.

Rosell, S., Kopin, I. J. & Axelrod, J. (1963). Fate of H^3-norepinephrine in skeletal muscle before and following nerve stimulation. *Am. J. Physiol.* **205**, 317–321.

Ross, S. B. (1963). *In vivo* inactivation of catecholamines in mice. *Acta pharmac. tox.* **20**, 267–273.

Ross, S. B. & Haljasmaa, Ö. (1964*a*). Catechol-*O*-methyl transferase inhibitors. *In vitro* inhibition of the enzyme in mouse-brain extract. *Acta pharmac. tox.* **21**, 205–214.

Ross, S. B. & Haljasmaa, Ö. (1964*b*). Catechol-*O*-methyl transferase inhibitors. *In vivo* inhibition in mice. *Acta pharmac. tox.* **21**, 215–225.

Sanan, S. & Vogt, M. (1962). Effect of drugs on the noradrenaline content of brain and peripheral tissues and its significance. *Br. J. Pharmac. Chemother.* **18**, 109–127.

Schmiterlöw, C. G. (1951). The formation *in vivo* of noradrenaline from 3,4-dihydroxyphenylserine (noradrenaline carboxylic acid). *Br. J. Pharmac. Chemother.* **6**, 127–134.

Schümann, H. J. (1958). Über die Verteilung von Noradrenalin und Hydroxytyramin in sympathischen Nerven (Milznerven). *Arch. exp. Path. Pharmak.* **234**, 17–25.

Sharman, D. F., Vanov, S. & Vogt, M. (1962). Noradrenaline content in the heart and spleen of the mouse under normal conditions and after administration of some drugs. *Br. J. Pharmac. Chemother.* **19**, 527–533.

Shore, P. A., Busfield, O. & Alpers, H. S. (1964). Binding and release of metaraminol: mechanism of norepinephrine depletion by α-methyl-*m*-tyrosine and related agents. *J. Pharmac. exp. Ther.* **146**, 194–199.

Snyder, S. H., Fischer, J. & Axelrod, J. (1965). Evidence for the presence of monoamine oxidase in sympathetic nerve endings. *Biochem. Pharmac.* **14**, 363–365.

Sourkes, T. L. (1954). Inhibition of dihydroxyphenylalanine decarboxylase by derivatives of phenylalanine. *Archs. Biochem. Biophys.* **51**, 444–456.

Sourkes, T. L. (1966). DOPA decarboxylase: substrates, coenzyme, inhibitors. (Second International Catecholamine Symposium, Milan.) *Pharmacol. Rev.* **18**, 53–60.

Spector, S., Maling, H. M. & Shore, P.A. (1959). The effect of JB 516, a monoamine oxidase inhibitor on levels of serotonin and norepinephrine in brain and spinal cord. *Fedn. Proc.* **18**, 447.

Spector, S., Mater, R. O., Sjoerdsma, A. & Udenfriend, S. (1965). Biochemical and pharmacological effects of iodo-tyrosines; relation to tyrosine hydroxylase inhibition *in vivo*. *Life Sciences* **4**, 1307–1311.

Spector, S., Melmon, K. & Sjoerdsma, A. (1962). Evidence for rapid turnover of norepinephrine in rat heart and brain. *Proc. Soc. exp. Biol. Med.* **111**, 79–81.

Spector, S., Shore, P. & Brodie, B. B. (1960). Biochemical and pharmacological effects of monoamine oxidase inhibitors, iproniazid, 1-phenyl-2-hydrazinopropane (JB 516) and 1-phenyl-3-hydrazinobutane. *J. Pharmac. exp. Ther.* **128**, 15–21.

Spector, S., Sjoerdsma, A. & Udenfriend, S. (1965). Blockade of endogenous norepinephrine synthesis by α-methyl-tyrosine, an inhibitor of tyrosine hydroxylase. *J. Pharmac. exp. Ther.* **147**, 86–95.

Spector, S., Sjoerdsma, A., Zaltman-Nirenberg, P., Levitt, M. & Udenfriend, S. (1963). Norepinephrine synthesis from tyrosine-C^{14} in isolated perfused guinea pig hearts. *Science* **139**, 1299–1301.

Stinson, R. H. (1961). Electrical stimulation of the sympathetic nerves of the isolated rabbit ear and the fate of the neurohormone released. *Canad. J. Biochem. Physiol.* **39**, 309–316.

Stjärne, L. (1966). Storage particles in noradrenergic tissues. (Second International Catecholamine Symposium, Milan.) *Pharmacol. Rev.* **18**, 425–432.

Udenfriend, S. (1966). Tyrosine hydroxylase. (Second International Catecholamine Symposium, Milan.) *Pharmacol. Rev.* **18**, 43–52.

Udenfriend, S. & Creveling, C. R. (1959). Localization of dopamine-β-oxidase in brain. *J. Neurochem.* **4**, 350–352.

Udenfriend, S. & Wyngaarden, J. B. (1956). Precursors of adrenal epinephrine and norepinephrine *in vivo*. *Biochim. Biophys. Acta* **20**, 48–52.

Udenfriend, S. & Zaltzman-Nirenberg, P. (1963). Norepinephrine and 3,4-dihydroxyphenylethylamine turnover in guinea pig brain *in vivo*. *Science* **142**, 394–396.

Udenfriend, S., Zaltzman-Nirenberg, P. & Nagatsu, T. (1965). Inhibitors of purified beef adrenal tyrosine hydroxylase. *Biochem. Pharmac.* **14**, 837–846.

Van der Schoot, J. B. & Creveling, C. R. (1965). Substrates and inhibitors of dopamine-β-hydroxylase. *Adv. Drug Res.* **2**, 47–88.

Waltman, S. & Sears, M. (1964). Catechol-*O*-methyl transferase and monoamine oxidase activity in the ocular tissues of albino rats. *Invest. Opthalm.* **3**, 601–605.

Weiss, P. (1961). The concept of perpetual neuronal growth and proximo-distal substance convection. In *Regional Neurochemistry* (ed. S. S. Kety, and J. Elkes), p. 240. New York: Pergamon Press.

Werle, E. & Aures, D. (1959). Über die Reinigung und Spezifität der DOPA-Decarboxylase. *Hoppe-Seyler's Z. physiol. Chem.* **316**, 45–60.

Whitby, L. G., Axelrod, J. & Weil-Malherbe, H. (1961). The fate of H^3-norepinephrine in animals. *J. Pharmac. exp. Ther.* **132**, 193–201.

Wurtman, R. J. & Axelrod, J. (1966). Control of enzymatic synthesis of adrenaline in the adrenal medulla by adrenal cortical steroids. *J. biol. Chem.* **241**, 2301–5.

Wylie, D. W., Archer, S. & Arnold, A. (1960). Augmentation of pharmacological properties of catecholamines by *O*-methyl transferase inhibitors. *J. Pharmac. exp. Ther.* 130, 239–244.

Zeller, E. A. (ed.) (1959). Amine oxidase inhibitors. *Ann. N.Y. Acad. Sci.* **80**, 551–1046.

Zeller, E. A. & Barsky, J. (1952). *In vivo* inhibition of liver and brain monoamine oxidase by 1-isonicotinyl-2-isopropyl hydrazine. *Proc. Soc. exp. Biol. Med.* **18**, 459–461.

6

THE RELEASE OF NORADRENALINE FROM ADRENERGIC NERVES

The stimulation of adrenergic nerves, either directly or reflexly, causes a release of noradrenaline from the terminal regions which then diffuses to adjacent effector cells. Because of the large number of discrete terminal ramifications of adrenergic nerves the diffusion distance involved is only very small (200–500 Å). It seems likely, as discussed in chapter 4, that each effector cell is in close contact with at least one adrenergic fibre, and possibly with several such fibres. The earlier theory of Cannon & Rosenblueth (1937), which postulated that only certain key cells in the effector tissue were directly innervated and that the response of other cells in the tissue was the consequence of transmitter diffusion over relatively long distances within the tissue, is no longer tenable in the light of modern anatomical findings. Little is known of the detailed mechanism by which the arrival of nerve impulses at the terminals of adrenergic nerves causes the release of noradrenaline, although the possibly analogous release of catecholamines from the adrenal medulla has been extensively studied (Douglas, 1966).

1. Experimental approaches

A variety of experimental approaches has been used to study the release of the adrenergic transmitter in response to nerve stimulation. Much of the present knowledge of the physiology of adrenergic transmission was derived from indirect studies of this problem, in which the release of transmitter in response to nerve stimulation was studied by measurements of its effects on effector tissues. This approach had the advantage that the release mechanism could be studied in a semiquantitative manner. In this way the well-known spatial and temporal summation phenomena characteristic of sympathetically innervated tissues were described by Cannon & Rosenblueth (1937) and Rosenblueth (1950). These authors also devised techniques for studying the release of noradrenaline from stimulated organs by observing the effects of the released transmitter on some remote target organ. This technique was used by Cannon & Uridil (1921) in their original demonstration of the release of a chemical transmitter from sympathetic nerves. The effects of the small amounts of noradrenaline overflowing into the blood from a stimulated organ were detected on remote organs which had been made supersensitive to the actions of catecholamines by denervation or by cocaine treatment. Using this method it was possible to detect transmitter release not only by direct stimulation of sympathetic nerves but also by reflex stimulation (Liu & Rosenblueth, 1935).

Much of the direct evidence for the role of noradrenaline as the adrenergic transmitter came from experiments in which noradrenaline was detected in the venous effluent of organs such as the heart, liver or spleen during direct stimulation of their sympathetic nerve supply (chapter 3). With improved methods for the isolation and assay of noradrenaline such methods can be used for precise studies of the release of noradrenaline (Peart, 1949; Mirkin & Bonnycastle, 1954; Brown & Gillespie, 1957). A further advance has been to study the release of noradrenaline in isolated perfused organs subject to direct stimulation of their sympathetic nerves. This was the approach first used by Loewi (1921) in his classical demonstration of the release of adrenaline in the isolated frog heart. Similar studies have been performed with a variety of isolated organs: intestine (Finkleman, 1930); perfused rabbit ear (Gaddum & Kwiatkowski, 1939); ox spleen (Eliasson, Euler & Stjärne, 1955). It is only recently, however, that precise quantitative studies of the release of noradrenaline in response to nerve stimulation have been made. Such studies have been performed on the release of noradrenaline in the isolated cat spleen and in the intestine by Brown and coworkers (Brown, 1960, 1965). The findings of Brown's group have subsequently been confirmed by several workers—notably by Thoenen, Hürlimann & Haefely (1963, 1964*a*) using the cat spleen preparation.

A useful innovation has been to label the noradrenaline store in adrenergic nerve terminals by the introduction of small amounts of radioactively labelled noradrenaline. The small amounts of labelled noradrenaline are of sufficiently high specific activity to allow the subsequent detection of labelled noradrenaline released in response to nerve stimulation (Hertting & Axelrod, 1961; Rosell, Kopin & Axelrod, 1963; Hertting, 1965; Gillespie & Kirpekar, 1965).

2. Influence of stimulus frequency on noradrenaline release

An analysis of the effects produced by different frequencies of stimulation of sympathetic nerves in various tissues has shown that the adrenergic effector tissues respond sensitively to very low frequencies of stimulation. Maximum responses of effector tissues can usually be obtained with stimulus frequencies of less than 20/sec (Cannon & Rosenblueth, 1937). The relation between response and stimulation frequency has the shape of a rectangular hyperbole, thus very great quantitative differences in response may be elicited by minute changes in stimulation frequency (Fig. 6.1). Most tissues respond most sensitively to frequency changes in the range 1–3/sec. These findings correlate well with those of Folkow (1952), Celander (1954) and Mellander (1960) who concluded from extensive *in vivo* studies that the normal rate of discharge of postganglionic sympathetic fibres is extremely low. The normal rate of firing rarely exceeds 10/sec and is generally in the range 1–2/sec. These results suggest that adrenergic synapses normally operate at very low frequencies of stimulation. In the cat spleen the highest amounts of noradrenaline released per

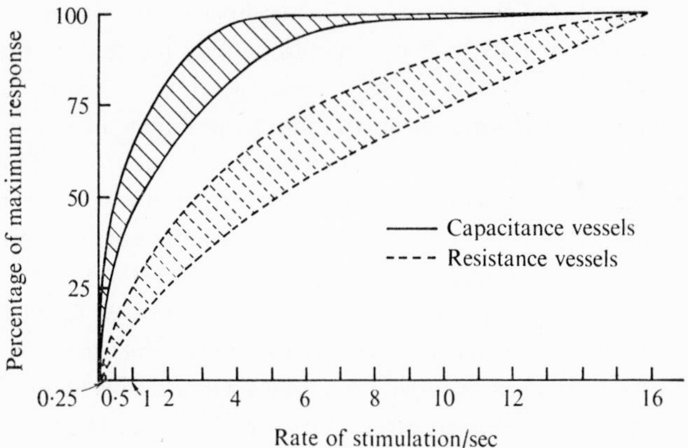

Fig. 6.1. The correlation between the rate of stimulation of adrenergic vasoconstrictor fibres and the response of resistance and capacitance blood vessels. Shaded areas indicate variations in the results obtained from a series of forty cats. From Mellander (1960).

stimulus were found at the lowest frequencies tested (10/sec) and declined as the stimulation frequency was increased to 30/sec or more (Brown & Gillespie, 1957). The amount of noradrenaline released per stimulus could only be measured in these experiments after the interfering factor of transmitter inactivation in the tissue had been eliminated; the rate of transmitter release should not be confused with the rate of transmitter 'overflow' at different frequencies. In similar experiments in the cat spleen Haefely, Hürlimann & Thoenen (1965) found that the amount of noradrenaline liberated per stimulus was constant at frequencies of 1–8/sec, but was considerably less at very low frequencies (0·5/sec).

Under normal physiological conditions there may also be a small but significant spontaneous release of noradrenaline from adrenergic nerve terminals in the absence of nerve impulses. In the perfused rabbit heart Huković & Muscholl (1962) found a basal rate of noradrenaline release of 2 ng/min, and this was increased more than 100-fold on stimulation of the sympathetic nerve supply. Burn & Rand (1958) observed that the spontaneous rate of isolated rabbit atria was significantly lower in atria in which the noradrenaline stores had been depleted with reserpine than in atria from normal animals. They suggested that the spontaneous release of noradrenaline in normal atria might influence the rate of beating. This suggestion was supported by the finding that in normal atria the rate of beating was increased if the inactivation of noradrenaline was inhibited by the addition of cocaine to the suspending medium (Huković, 1959). The demonstration of small spontaneous depolarizing potentials in sympathetically innervated smooth muscle by Burnstock & Holman (1962) also supports the view that a slow spontaneous release of transmitter may occur. The occurrence of spontaneous miniature junction potentials in smooth muscle is also reminiscent of the occurrence of miniature end-plate potentials in skeletal muscle, which indicate a spontaneous release of acetylcholine from cholinergic synapses (Katz, 1962). It is possible that the spontaneous release of small amounts of noradrenaline may play some role in the regulation of vasomotor tone.

3. Transmitter overflow

The success of the early experiments which demonstrated the release of noradrenaline by its remote effects on other organs, or by its presence in the venous effluent or perfusate outflow showed that under certain conditions the noradrenaline released by nerve stimulation accumulates in the effector tissue and overflows into the circulation. However, most of these experiments involved very high stimulation frequencies. It seems likely that very little overflow of noradrenaline occurs under normal physiological conditions, with stimulation frequencies of the order 1–2/sec. The experiments of Brown's group (Brown & Gillespie, 1957; Brown, Davies & Gillespie, 1958; Brown, Davies & Ferry, 1961; Brown, 1960, 1965) on noradrenaline release in the cat spleen and intestine have shown that at low frequencies of stimulation only very little of the released transmitter overflows and appears in the perfusate outflow (Fig. 6.2). Such overflow only becomes important at high stimulation frequencies, with a maximum overflow per stimulus at a frequency of 30/sec. The results obtained were quite different if the process of transmitter inactivation was inhibited by drug treatment; it was found that in the presence of certain adrenergic blocking drugs (dibenzyline, phentolamine, hydergine) the rate of noradrenaline output was very much higher at low frequencies of stimulation, although it did not then increase further as stimulation frequencies were increased to 30/sec (Fig. 6.2). These results indicate that at low frequencies of stimulation as much as 90 % of the released noradrenaline was inactivated by some local process within the effector tissue and did not overflow into the venous effluent. These results have subsequently been confirmed in similar experiments in the cat nictitating membrane and spleen and in the rabbit eye (Haefely, Hürlimann & Thoenen, 1964; Thoenen, Hürlimann & Haefely, 1964*b*; Eakins & Eakins, 1964). At physiological frequencies of stimulation only small amounts of noradrenaline overflow into the circulation. This conclusion was also reached by Celander (1954), who failed to detect any remote effects on the supersensitive denervated nictitating membrane even after stimulation of certain sympathetic nerves at high frequencies. Celander also found that catecholamines were removed very rapidly from circulation after their infusion. In the

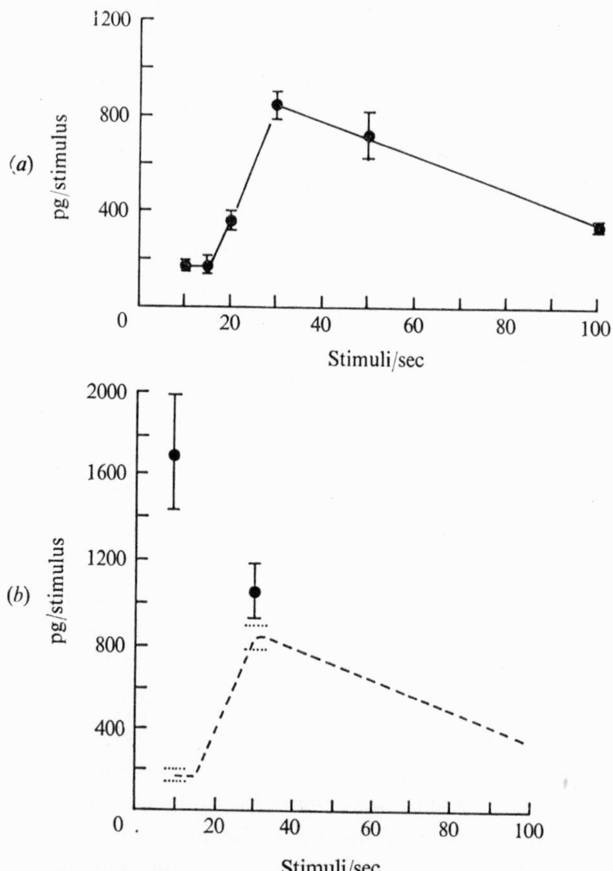

Fig. 6.2. Overflow of noradrenaline into venous blood of cat spleen after stimulation of the splenic nerve at various frequencies. (*a*) Mean values, with standard errors of the mean, for a series of experiments in normal animals. (*b*) Values obtained after treatment with adrenergic blocking agents (phenoxybenzamine or hydergine); stimulation of splenic nerve at 10/sec (8 expts.) or 30/sec (14 expts.). Normal outflow curve is indicated by dotted line. (From Brown, 1965.)

light of these later findings it seems that the success of the early experiments of Cannon & Rosenblueth may have been due to a fortuitous combination of experimental conditions.

Despite the relatively minor amounts of transmitter overflowing into the circulation from adrenergic synapses, relatively large amounts of noradrenaline metabolites are excreted in the urine and appreciable levels of noradrenaline are found in plasma (chapter 4). In man the daily urinary excretion of catecholamine metabolites amounts to the equivalent of some 5–10 mg of noradrenaline. It seems that the majority of these products derive from neuronally released noradrenaline, since the urinary excretion of noradrenaline and its metabolites is not significantly lowered after adrenalectomy (Euler, Franksson & Hellström, 1954). Furthermore, increased adrenergic activity is well correlated with rises in the urinary excretion of noradrenaline (Euler, Björkman & Orwén, 1955) and ganglionic blocking agents markedly decrease the urinary excretion of noradrenaline (Sundin, 1958). Of the total urinary catecholamine metabolities, only about 20 % derive from adrenaline released from the adrenal medulla. From experiments which measured the rate of urinary excretion of noradrenaline after various rates of intravenous infusion, Euler & Luft (1951) calculated that in normal human subjects 0·5–1·0 μg of noradrenaline per minute overflow into the circulation.

It seems probable that appreciable amounts of released transmitter do overflow into the circulation from peripheral adrenergic synapses, but the amounts of noradrenaline thus removed and subsequently excreted probably represent only a small proportion of the total amount released from nerve endings, most of which is inactivated locally in peripheral tissues. This conclusion is supported by the finding that the urinary excretion of noradrenaline is greatly increased when the tissue inactivation process is inhibited with drugs such as dibenzyline (Benfey, Ledoux & Melville, 1959).

4. Possible mechanisms of release

Very little is known of the mechanism by which nerve impulses promote the release of noradrenaline from adrenergic nerves. In the adrenal medulla the acetylcholine-mediated release of catecholamines involves an entry of calcium ions into the medullary cells. Thus the amount of catecholamine released from the perfused cat adrenal in response to acetylcholine was dependent on the calcium content of the perfusing medium (Douglas & Rubin, 1961). Moreover, under certain conditions calcium alone provoked a secretory response. An enhanced uptake of Ca^{45} by adrenal medulla tissue in the presence of acetylcholine has also been demonstrated (Douglas & Poisner, 1961, 1962). Calcium has also been shown to release catecholamines from isolated medullary storage particles incubated *in vitro* (Schümann & Philippu, 1963). The release of noradrenaline from the isolated perfused rabbit heart in response to sympathetic nerve stimulation was considerably reduced if the calcium content of the perfusing medium was lowered to less than 1 m-equiv./l. (Huković & Muscholl, 1962). Similarly, the release of H^3-noradrenaline in response to stimulation of the sympathetic nerve supply to the isolated cat colon is greatly reduced in calcium-free medium (Boullin, 1965). The response of the rabbit ileum to stimulation of the sympathetic nerve supply is also greatly reduced or absent in calcium-free medium (Burn & Gibbons, 1965). It seems, therefore, that the release of noradrenaline from sympathetic nerves in response to nerve impulses is also a calcium-dependent process.

Despite the sparsity of information concerning the mechanism of noradrenaline release at adrenergic synapses, the process seems to be analogous in some respects to the more extensively studied release of acetylcholine at cholinergic synapses (Eccles, 1964). Both noradrenaline and acetylcholine are stored in submicroscopic vesicles in the presynaptic nerve terminals. Burnstock & Holman (1962) have recorded a spontaneous discharge of small potentials in sympathetically innervated smooth-muscle cells of the guinea-pig vas deferens (Fig. 6.3). This may imply that noradrenaline is released in quantal packets, in the same way as acetylcholine (Katz, 1962; for review see Burnstock & Holman, 1963). The output of noradrenaline per stimulus decreases with increasing stimulation frequencies (Brown & Gillespie, 1957) in a manner qualitatively similar to the release of acetylcholine. Both processes are also dependent on calcium ions and in each case transmitter release is promoted by depolarizing concentrations of extracellular potassium (Douglas & Rubin, 1963). It seems possible that the mechanism of noradrenaline release may prove to be similar to that suggested for acetylcholine by Katz (1962). The primary event that leads to transmitter release is a depolarization of the presynaptic terminal during the action potential of the invading nerve impulse. Katz has suggested that this depolarization may be followed by an increase in the number of attachment sites on the inner surface of the presynaptic fibre for the preformed packets of transmitter (storage vesicles ?); the transmitter packets attach to these sites and discharge their contents to the exterior into the synaptic cleft. Calcium is an essential cofactor for the latter stages. Recent findings from studies of the release of catecholamines from the adrenal medulla provide direct evidence that in this case the release involves a discharge of intracellular storage par-

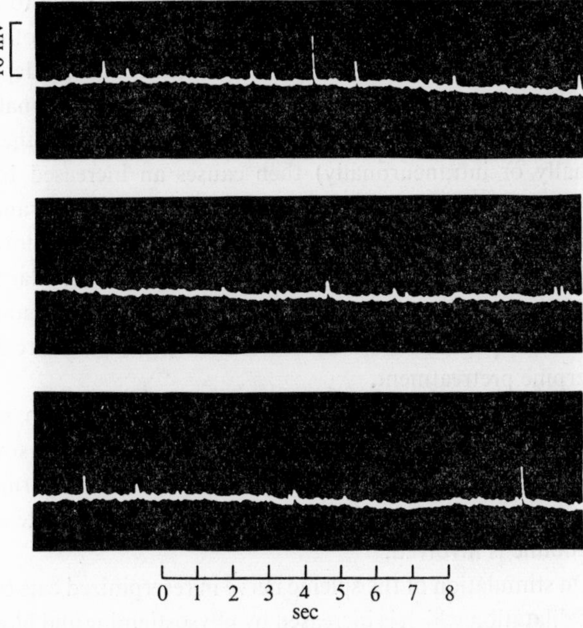

Fig. 6.3. Continuous record of intracellular membrane potential in smooth muscle fibres of guinea-pig vas deferens, showing a spontaneous discharge of 'miniature junction potentials' in the absence of nerve stimulation. (From Burnstock & Holman, 1962.)

ticles, or at least their contents. The secretion of catecholamines from the adrenal medulla is accompanied by a release of adenine nucleotides which are known to be contained in the medullary catecholamine storage particles (Douglas, 1966). The release of medullary catecholamines is also accompanied by the release of a specific protein, which can be isolated from the medullary storage particles (Kirshner, Sage, Smith & Kirshner, 1966).

5. Role of acetylcholine in adrenergic transmission

The hypothesis of a cholinergic link in the sympathetic postganglionic adrenergic transmission proposed by Burn & Rand (1959, 1960) has stimulated a great deal of investigation. The evidence in favour of this hypothesis has been reviewed recently by Burn & Rand (1965) and will be discussed only briefly here. According to Burn & Rand postganglionic sympathetic nerves may contain both acetylcholine and noradrenaline. The primary event following depolarization of the nerve terminals is not a release of noradrenaline but a release of acetylcholine. The released acetylcholine (either extraneuronally or intraneuronally) then causes an increased influx of calcium ions which in turn release noradrenaline from intraneuronal vesicles. The evidence in favour of this hypothesis is as follows:

(*a*) Acetylcholine or nicotine-like drugs cause effects similar to those evoked by stimulation of the sympathetic nerves in many organs. These effects are not present if the tissue stores of noradrenaline are depleted by reserpine pretreatment.

(*b*) Adrenergic neurone blocking agents such as bretylium, xylocholine and guanethidine have chemical structures which are somewhat similar to that of acetylcholine; it is proposed that these drugs interfere with adrenergic transmission by interacting with the site at which acetylcholine is involved.

(*c*) On stimulation of the splenic nerve in reserpinized cats there is a splenic dilatation which is increased by physostigmine and blocked by atropine; acetylcholine appears in the splenic outflow. Similar findings are reported in the isolated rabbit ear, the cat nictitating membrane and heart.

(*d*) The acetylcholine content of the cat spleen (0·47 μg/g) falls markedly (to 0·1 μg/g) after section and degeneration of the splenic nerve, suggesting that postganglionic sympathetic fibres contain appreciable amounts of acetylcholine.

(*e*) Electron-microscopic studies of the spleen and other sympathetically innervated organs have failed to reveal two populations of terminals, one noradrenergic and one cholinergic. Rather there is one type of autonomic nerve terminal which contains two morphologically distinct vesicles, a granulated and a non-granulated type. It is supposed that this implies that the nerves store both acetylcholine (agranular vesicles) and noradrenaline (granular vesicles).

(*f*) The sympathetic responses evoked by acetylcholine can be blocked by bretylium, which is known to act at adrenergic synapses.

(*g*) Adrenergic transmission can be blocked by two drugs which are known to act on cholinergic mechanisms: to block acetylcholine synthesis (hemicholinium) and to impair acetylcholine release (botulinum toxin).

Despite the plausibility of this evidence, the cholinergic link hypothesis is not widely accepted. Some of the evidence which argues against the hypothesis can be summarized as follows:

(*a*) Ferry (1963) was able to show that in the cat spleen the close arterial injection of acetylcholine caused a vigorous antidromic discharge of the postganglionic sympathetic nerve fibres of the splenic nerve. This suggests that the sympathomimetic actions of acetylcholine, nicotine and nicotine-like drugs are indeed due to a stimulation of the postganglionic sympathetic nerves, but that this effect is quite nonspecific. In favour of this view it is known that acetylcholine can stimulate sensory nerves, although it is clear that acetylcholine is not involved as a transmitter in such nerves (Gray & Diamond, 1957). The actions of acetylcholine at such a non-specific preterminal site in both sensory nerves and in sympathetic nerves can be blocked by hexamethonium which also blocks the sympathomimetic actions of acetylcholine.

(*b*) On the other hand, hexamethonium has no effect on the response of the cat spleen to stimulation of the sympathetic nerve supply. Nor do potent anticholinesterases in any way prolong or potentiate the effects of splenic nerve stimulation or increase the amounts of noradrenaline liberated (Blakeley, Brown & Ferry, 1963).

(*c*) Histochemical studies have failed to demonstrate the presence of acetylcholinesterase in postganglionic sympathetic neurones, except in certain sympathetic ganglion cells which are involved in sudomotor functions (Sjöqvist, 1963; Hamberger, Norberg & Sjöqvist, 1965), although this remains a controversial point (Koelle, 1965).

Although the arguments against the cholinergic link hypothesis are formidable, they cannot be said to answer all the points in favour of this hypothesis. Until more evidence is available it would be wise to avoid a definite conclusion in favour of or against this idea. Whatever the final verdict we should be grateful to Burn & Rand for putting forward a hypothesis which has led to a re-examination of well-established dogmas. The cholinergic-link hypothesis has certainly generated a great deal of heat, but it has also produced some light.

6. Drugs which interfere with the release of noradrenaline

There has been much interest in recent years in the development of drugs which interfere with the normal release of noradrenaline from adrenergic nerves in response to nerve stimulation. The effect of such drugs is to diminish sympathetic tone and to prevent the hypertensive effects of sympathetic overactivity. The hypotensive effects of these drugs have proved useful clinically in the treatment of certain forms of hypertension. There are several ways in which drugs may interfere with the normal release of the adrenergic transmitter and these will now briefly be discussed.

(*a*) *Adrenergic neurone blocking drugs*

A group of drugs which interfere with the release of noradrenaline in response to sympathetic nerve stimulation have become important in the treatment of hyptertensive diseases. These drugs are described as adrenergic neurone blocking agents to distinguish them from adrenergic receptor blocking or sympatholytic agents, which antagonize the actions of the released transmitter on postsynaptic receptors, or ganglionic blocking agents, which prevent the transmission of impulses from preganglionic to postganglionic sympathetic nerves. The pharmacology of the adrenergic neurone blocking agents has been reviewed recently by Boura & Green (1965) and by Copp (1964). A large number of substances has been found to possess the ability to prevent the normal release of noradrenaline from adrenergic nerves. Some well-known examples of the group are illustrated in Fig. 6.4. All active compounds have a basically similar structure: a highly basic unit (quaternary amine cationic head, guanidines, amidines, amidoximes) linked by a one- or two-carbon chain to a ring.

Adrenergic neurone blocking agents, such as xylocholine, bretylium and guanethidine, block the responses of various pharmacological test systems to adrenergic stimulation while the responses to exogenously administered noradrenaline remain unimpaired or potentiated. Direct evidence has shown that these compounds prevent the release of noradrenaline from the cat spleen on stimulation of the

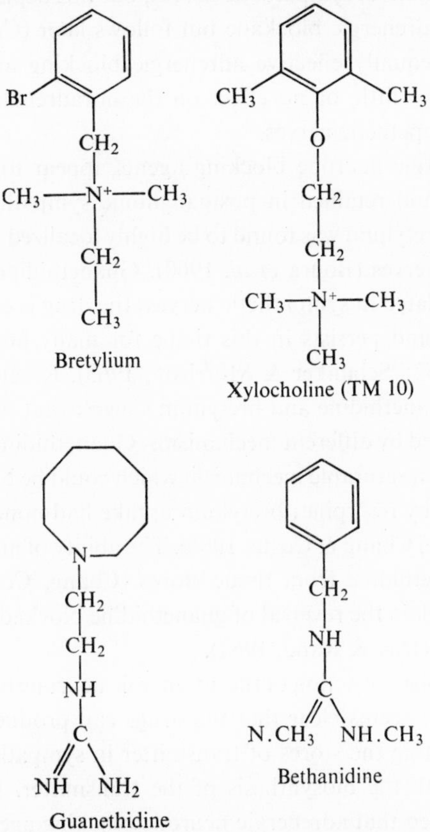

Bretylium

Xylocholine (TM 10)

Guanethidine

Bethanidine

Fig. 6.4. Structures of some adrenergic neurone blocking agents.

splenic nerve (Exley, 1957; Boura & Green, 1959, 1963; Abercrombie & Davies, 1963). While xylocholine (TM 10) shows muscarinic, nicotinic, ganglion-blocking and neuromuscular paralytic actions at doses only slightly higher than those needed to block adrenergic transmission, other adrenergic neurone blocking agents have little or no action on cholinergic mechanisms. Most adrenergic blocking agents produce transient sympathomimetic effects after administration, particularly at high doses. This effect seems to be due to an initial release of noradrenaline from the adrenergic nerve terminals. Guanethidine produces a long-lasting and almost total depletion of the noradrenaline content of peripheral sympathetic nerves, but this depletion is not the cause of the adrenergic blockade but follows later (Cass & Spriggs, 1961). Other equally effective adrenergic blocking agents, such as bretylium, have little or no effect on the noradrenaline content of peripheral sympethetic nerves.

The adrenergic neurone blocking agents appear to be selectively accumulated and retained in postganglionic sympathetic neurones. Radioactive bretylium was found to be highly localized in sympathetic ganglia and nerves (Boura *et al.* 1960). Guanethidine is also selectively accumulated in sympathetic nerves; the drug is concentrated in the rat heart and persists in this tissue for many hours (Bisson & Muscholl, 1962; Schanker & Morrison, 1965). Recent studies with radioactive guanethidine and bretylium suggest that these two drugs are accumulated by different mechanisms. Guanethidine was taken up in the heart by a saturable mechanism which could be blocked by noradrenaline or by reserpine; bretylium uptake had none of these properties (Brodie, Chang & Costa, 1965). The ability of amphetamine to displace guanethidine from tissue stores (Chang, Costa & Brodie, 1965) may explain the reversal of guanethidine blockade produced by amphetamine (Day & Rand, 1962).

The mechanism of action of the adrenergic neurone blocking agents is uncertain. It seems clear that the drugs can produce their effects without depleting the stores of transmitter in sympathetic nerves or interfering with the biosynthesis of the transmitter. Burn & Rand (1965) suggested that adrenergic neurone blocking agents may interfere with the postulated cholinergic link. A more probable explanation is that these agents act by preventing the propagation of nerve impulses in the fine fibres of the autonomic ground plexus. Although the adrenergic neurone blocking agents do not prevent the transmission of impulses in postganglionic sympathetic nerve trunks (Exley, 1957, 1960), many of these drugs have powerful and persistent local anaesthetic actions (Hey & Willey, 1954; Boura & Green, 1959). This property, together with the selective accumulation of high concentrations of these drugs in adrenergic terminals may explain their specific actions on adrenergic transmission. Burnstock & Holman (1965) reported that bretylium failed to produce a long-lasting block of the spontaneous occurrence of miniature potentials in the smooth muscle of the guinea-pig vas deferens, although the tissue no longer responded to stimulation of the sympathetic nerves. This is in accord with the view that the adrenergic neurone blocking agents do not act on the noradrenaline release mechanism itself, but block conduction in preterminal sympathetic fibres.

(*b*) *False adrenergic neurotransmitters*

Although noradrenaline is normally the only adrenergic transmitter in mammals it is now clear that under abnormal circumstances several other structurally related amines can be released from sympathetic nerve terminals by nerve stimulation. Such compounds, which can be stored in place of noradrenaline in adrenergic nerve terminals and released by nerve stimulation have been termed 'false transmitters'. The false transmitters may or may not stimulate postsynaptic effector sites after their release from nerve endings. If the false transmitter lacks physiological activity or is less potent than noradrenaline the net result of replacing the normal transmitter with a false transmitter will be to reduce sympathetic tone; any drugs which can lead to this type of situation may thus be clinically useful.

After the administration of α-methyl DOPA the long-lasting depletion of noradrenaline from peripheral sympathetic nerves is accompanied by an accumulation of α-methyl noradrenaline in these tissues (Carlsson & Lindqvist, 1962; Maître & Staehlin, 1963). Carlsson & Lindqvist (1962) and Day & Rand (1963, 1964) expressed the view that α-methyl noradrenaline might act as a false transmitter in adrenergic nerves. The latter authors demonstrated that α-methyl DOPA can restore the actions of sympathetic nerve stimulation in peripheral adrenergic nerves depleted of noradrenaline by reserpine treatment. α-Methyl DOPA is unable to restore adrenergic transmission in these animals if an inhibitor of dopamine-β-hydroxylase is given (Disulfuram), suggesting that the false transmitter formed from α-methyl DOPA is α-methyl noradrenaline (I. J. Kopin, personal communication). The first direct evidence of a release of α-methyl noradrenaline from the rabbit heart by nerve stimulation was provided by Muscholl &

Maître (1963). The depletion of tissue noradrenaline stores after α-methyl-*m*-tyrosine is similarly accompanied by an accumulation of metaraminol in peripheral tissues (Fig. 6.5) (Gessa, Costa, Kuntzman & Brodie, 1962; Andén, 1964). As shown by Shore, Busfield & Alpers (1964), the administration of small doses of metaraminol also leads to an accumulation of this amine in adrenergic nerves together

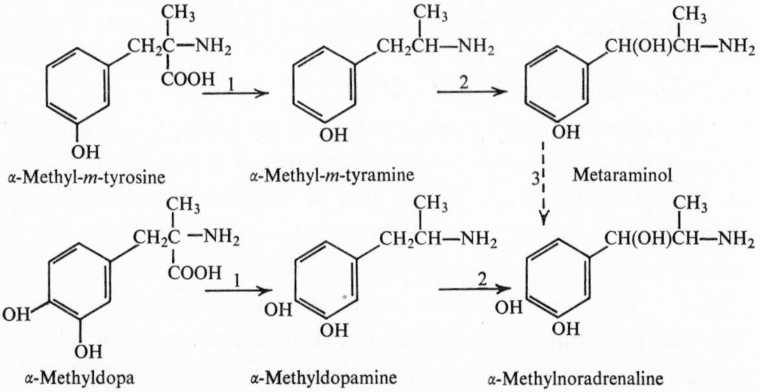

α-Methyl-*m*-tyrosine → (1) α-Methyl-*m*-tyramine → (2) Metaraminol

α-Methyldopa → (1) α-Methyldopamine → (2) α-Methylnoradrenaline

Metaraminol → (3) α-Methylnoradrenaline

Fig. 6.5. Metabolism of α-methyl amino acids to 'false transmitters'. 1 = aromatic-L-amino acid decarboxylase; 2 = dopamine-β-hydroxylase; 3 = liver microsomal hydroxylating enzyme.

with a depletion of the noradrenaline content. Metaraminol persists for long periods in the tissue stores and is said to replace noradrenaline stoichiometrically (Shore *et al.* 1964; Andén, 1964), although Udenfriend & Zaltzman-Nirenberg (1964) reported that the noradrenaline deficit in metaraminol-depleted guinea-pig hearts was not accounted for by an equivalent amount of metaraminol. However, metaraminol itself can be converted to α-methyl noradrenaline *in vivo* (Maître & Staehlin, 1965) and it is possible that the sum of metaraminol + α-methyl noradrenaline can account for the noradrenaline deficit. Crout & Shore (1964) demonstrated that metaraminol too could be released by nerve stimulation in the isolated guinea-pig heart.

Although α-methyl DOPA and α-methyl-*m*-tyrosine have proved to have antihypertensive effects in man, evidence that these drugs suppress adrenergic function in experimental animals is conflicting, probably because of species differences in the sensitivity of adrenergic receptors to the false transmitters α-methyl noradrenaline or metaraminol. Several authors have reported that adrenergic transmission is unaffected by α-methyl-*m*-tyrosine (Andén, 1964; Carlsson, 1964), but Haefely, Thoenen & Hürlimann (1965) reported that treatment with this compound considerably reduced the effects of sympathetic nerve stimulation in the cat nictitating membrane and spleen. In the cat spleen the output of noradrenaline on nerve stimulation was markedly reduced. For reviews of the pharmacology of the α-methyl amino acids see Stone & Porter (1966) and Muscholl (1966).

Adrenaline, although not normally present in sympathetic nerves, can be taken up by sympathetic nerve endings and released in response to nerve stimulation (Rosell, Kopin & Axelrod, 1964). In addition other β-hydroxylated sympathomimetic amines such as octopamine and α-methyl octopamine can be taken up and released from sympathetic nerves in the cat spleen (Fig. 6.6) (Fischer, Horst & Kopin, 1965); other β-hydroxylated amines which could be released were phenylethanolamine and norephedrine. The non-β-hydroxylated amines, tyramine, α-methyl tyramine, phenylethylamine and amphetamine were not released as false transmitters in the spleen. The authors point out the correlation between these results and those of Musacchio, Fischer & Kopin (1965) and Musacchio, Kopin & Weise (1965) who showed that the β-hydroxylated amines are retained in adrenergic nerve storage particles in the rat heart whereas the non-β-hydroxylated amines were not retained. The exception is dopamine and its α-methyl derivative, which are both retained in storage granules and are released as false transmitters (J. E. Fischer, personal communication). These results suggest that in order to be released from adrenergic terminals a substance must first be stored in intraneuronal particles, implying that the granules are intimately connected with the transmitter release mechanism.

Following the finding that several β-hydroxylated amines can act as false adrenergic transmitters Kopin, Fischer Musacchio & Horst (1964) and Kopin *et al.* (1965) have proposed an ingenious theory to explain the sympathetic blockade and hypotensive effects produced by the prolonged administration of monoamine oxidase inhibitors. After the prolonged administration of such drugs the release of noradrenaline from adrenergic nerves is impaired (Davey, Farmer &

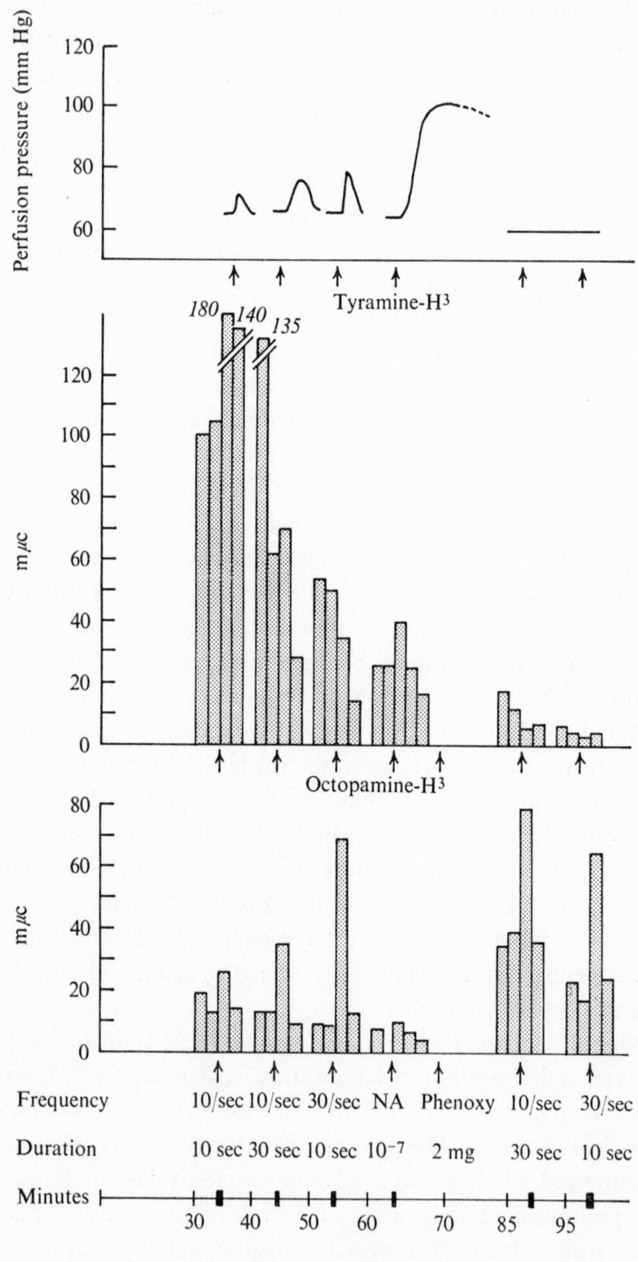
Perfusion pressure (mm Hg)
120
100
80
60
Tyramine-H3
180
140
135
mμc
120
100
80
60
40
20
0
Octopamine-H3
80
60
40
20
0
Frequency
10/sec 10/sec 30/sec NA Phenoxy 10/sec 30/sec
Duration
10 sec 30 sec 10 sec 10−7 2 mg 30 sec 10 sec
Minutes
30 40 50 60 70 85 95

Reinert, 1963) and there is an increased urinary excretion of amines such as tyramine which are normally destroyed by monoamine oxidase (Sjoerdsma *et al.* 1959). Under such conditions considerable amounts of the β-hydroxylated metabolite of tyramine, octopamine, accumulate in peripheral tissues (Kakimoto & Armstrong, 1962). Kopin *et al.* (1964, 1965) were able to show that endogenous octopamine accumulated only in tissues with an intact sympathetic innervation, indicating that the amine accumulates in sympathetic nerves where it is then released with noradrenaline and has the effect of diluting the normal transmitter with a much less potent substance. The net effect is to reduce the response of the innervated tissues to nerve activity.

(*c*) *Other drugs*

In addition to drugs which depress adrenergic transmission by preventing the release of the normal transmitter or by replacing the normal transmitter with a less active false transmitter, adrenergic function can be depressed in other ways. Of major importance in this context are the drugs of the Rauwolfia alkaloid group, of which reserpine is the best known example. Reserpine produces a long lasting block of adrenergic transmission and a profound depletion of the noradrenaline content of sympathetic nerves (Carlsson, Rosengren, Bertler & Nilsson, 1957; Holzbauer & Vogt, 1956; Muscholl & Vogt, 1958). In this case the principle action of the drug appears to be to prevent the normal storage of noradrenaline in intraneuronal granules (see chapter 8).

Fig. 6.6. Release of octopamine as a false transmitter in the cat spleen. Release of H^3-tyramine (upper histogram) and H^3-octopamine (lower histogram) from the isolated perfused spleen of the cat after infusion of H^3-tyramine (50 μc) at zero time. As indicated by arrows, splenic nerve was stimulated at 10 or 30/sec. While the outflow of H^3-tyramine appeared to be unrelated to nerve stimulation, the release of H^3-octopamine was clearly enhanced by such stimulation. The release of H^3-octopamine was unrelated to the contraction of the spleen produced by nerve stimulation since contractions induced by the intra-arterial injection of noradrenaline (10 ng, NA) failed to release H^3-octopamine, and octopamine release in response to nerve stimulation persisted after contraction was blocked by the administration of phenoxybenzamine (Phenoxy). Results are typical of six experiments; each bar represents a 2 min collection period. (From Fischer, Horst & Kopin, 1965.)

Another drug which produces a block in adrenergic transmission together with a long lasting depletion of the noradrenaline stores is 6-hydroxydopamine (2,4,5-trihydroxyphenylethylamine) (Porter, Totaro & Stone, 1963; Laverty, Sharman & Vogt, 1965). A final group of drugs which can block adrenergic transmission are the recently discovered inhibitors of noradrenaline biosynthesis, such as α-methyl-*p*-tyrosine, which lead to a failure in adrenergic transmission when the available stores of transmitter are exhausted (chapter 5).

REFERENCES

Abercrombie, G. F. & Davies, B. N. (1963). The action of guanethidine with particular reference to the sympathetic nervous system. *Br. J. Pharmac. Chemother.* **20**, 171–177.

Andén, N. E. (1964). On the mechanism of noradrenaline depletion by α-methylmetatyrosine and metaraminol. *Acta Pharmac. tox.* **21**, 260–271.

Benfey, B. G., Ledoux, G. & Melville, K. I. (1959). Increased urinary excretion of adrenaline and noradrenaline after phenoxybenzamine. *Br. J. Pharmac. Chemother.* **14**, 142–148.

Bisson, G. M. & Muscholl, E. (1962). Die Beziehung zwischen der Guanethidin-Konzentration im Rattenherzen und dem Noradrenalingehalt. *Arch. exp. Path. Pharmak.* **244**, 185–194.

Blakeley, A. G. H., Brown, G. L. & Ferry, C. B. (1963). Pharmacological experiments on the release of the sympathetic transmitter. *J. Physiol., Lond.* **167**, 505–514.

Boullin, D. J. (1965). Effect of divalent ions on release of ^{3}H-noradrenaline by sympathetic nerve stimulation. *J. Physiol., Lond.* **183**, 28–29*P*.

Boura, A. L. A., Copp, F. C., Duncombe, W. G., Green, A. F. & McCoubrey, A. (1960). The selective accumulation of bretylium in sympathetic ganglia and their postganglionic nerves. *Br. J. Pharmac. Chemother.* **15**, 265–270.

Boura, A. L. A. & Green, A. F. (1959). The actions of bretylium, adrenergic neurone blocking and other effects. *Br. J. Pharmac. Chemother.* **14**, 536–548.

Boura, A. L. A. & Green, A. F. (1963). Adrenergic neurone blockade and other acute effects caused by *N*-benzyl-*N′N″*-dimethylguanidine and its ortho-chloro derivative. *Br. J. Pharmac. Chemother.* **20**, 36–55.

Boura, A. L. A. & Green, A. F. (1965). Adrenergic neurone blocking agents. *A. Rev. Pharmac.* **5**, 183–212.

Brodie, B. B., Chang, C. C. & Costa, E. (1965). On the mechanism of action of guanethidine and bretylium. *Br. J. Pharmac. Chemother.* **25**, 171–178.

Brown, G. L. (1960). Release of sympathetic transmitter by nerve stimulation. In *Adrenergic Mechanisms*, pp. 116–124. CIBA. London: Churchill.

Brown, G. L. (1965). The Croonian Lecture 1964. The release and fate of the transmitter liberated by adrenergic nerves. *Proc. R. Soc.* B, **162**, 1–19.

Brown, G. L., Davies, B. M. & Ferry, C. B. (1961). The effect of neuronal rest on the output of sympathetic transmitter from the spleen. *J. Physiol., Lond.* **159**, 365–380.

Brown, G. L., Davies, B. M. & Gillespie, J. S. (1958). The release of chemical transmitter from the sympathetic nerves of the intestine of the cat. *J. Physiol., Lond.* **143**, 41–54.

Brown, G. L. & Gillespie, J. S. (1957). The output of sympathetic transmitter from the spleen of the cat. *J. Physiol., Lond.* **138**, 81–102.

Burn, J. H. & Gibbons, W. R. (1965). The release of noradrenaline from sympathetic fibres in relation to calcium concentration. *J. Physiol., Lond.* **181**, 214–223.

Burn, J. H. & Rand, M. J. (1958). Action of nicotine on the heart. *Br. med. J.* **1**, 137–139.

Burn, J. H. & Rand, M. J. (1959). Sympathetic postganglionic mechanism. *Nature, Lond.* **184**, 163–165.

Burn, J. H. & Rand, M. J. (1960). Sympathetic postganglionic cholinergic fibres. *Br. J. Pharmac. Chemother.* **15**, 56–66.

Burn, J. H. & Rand, M. J. (1965). Acetylcholine in adrenergic transmission. *A. Rev. Pharmac.* **5**, 163–182.

Burnstock, G. & Holman, M. E. (1962). Spontaneous potentials at sympathetic nerve endings in smooth muscle. *J. Physiol., Lond.* **160**, 446–460.

Burnstock, G. & Holman, M. E. (1963). Smooth muscle: autonomic nerve transmission. *A. Rev. Physiol.* **25**, 61–90.

Burnstock, G. & Holman, M. E. (1965). An electrophysiological investigation of the actions of some autonomic blocking drugs on transmission in the guinea pig vas deferens. *Br. J. Pharmac. Chemother.* **23**, 600–612.

Cannon, W. B. & Rosenblueth, A. (1937). *Autonomic Neuroeffector Systems.* New York: Macmillan.

Cannon, W. B. & Uridil, J. E., (1921). Studies on the conditions of activity in endocrine glands. VIII. Some effects on the denervated heart of stimulating the nerves of the liver. *Am. J. Physiol.* **58**, 353–354.

Carlsson, A. (1964). Functional significance of drug induced changes in brain monoamine levels. *Prog. Brain Res.* **8**, 9–27.

Carlsson, A. & Lindqvist, M. (1962). *In vivo* decarboxylation of α-methyl dopa and α-methyl-*m*-tyrosine. *Acta physiol. scand.* **54**, 87–94.

Carlsson, A., Rosengren, E., Bertler, A. & Nilsson, J. (1957). Effect of reserpine on the metabolism of catecholamines. In *Psychotropic Drugs*, ed. Garattini and Ghetti. Amsterdam: Elsevier.

Cass, R. & Spriggs, T. L. B. (1961). Tissue amine levels and sympathetic blockade after guanethidine and bretylium. *Br. J. Pharmac. Chemother.* **17**, 442–450.

Celander, O. (1954). The range of control exercised by the sympathico-adrenal system. *Acta physiol. scand.* **32**, Suppl. 116.

Chang, C. C., Costa, E. & Brodie, B. B. (1965). Interaction of guanethidine with adrenergic neurones. *J. Pharmac. exp. Ther.* **147**, 303–312.

Copp, F. C. (1964). Adrenergic neurone blocking agents. *Adv. Drug Res.* **1**, 167–189.

Crout, J. R. & Shore, P. A. (1964). Release of metaraminol (aramine) from the heart by sympathetic nerve stimulation. *Clin. Res.* **12**, 180.

Davey, M. J., Farmer, J. B. & Reinert, H. (1963). The effects of nialamide on adrenergic functions. *Br. J. Pharmac. Chemother.* **20**, 121–134.

Day, M. D. & Rand, M. J. (1962). Antagonism of guanethidine by dexamphetamine and other related sympathomimetic amines. *J. Pharm. Pharmac.* **14**, 541–549.

Day, M. D. & Rand, M. J. (1963). A hypothesis for the mode of action of α-methyldopa in relieving hypertension. *J. Pharm. Pharmac.* **15**, 221–224.

Day, M. D. & Rand, M. J. (1964). Some observations on the pharmacology of α-methyl dopa. *Br. J. Pharmac. Chemother.* **22**, 72–86.

Douglas, W. W. (1966). The mechanism of the release of catecholamines from adrenal medulla. (Second International Catecholamine Symposium.) *Pharmac. Rev.* **18**, 471–480.

Douglas, W. W. & Poisner, A. M. (1961). Stimulation of uptake of calcium^{45} in the adrenal gland by acetylcholine. *Nature, Lond.* **192**, 1299.

Douglas, W. W. & Poisner, A. M. (1962). On the mode of action of acetylcholine in evoking adrenal medullary secretion: increased uptake of calcium during the secretory response. *J. Physiol., Lond.* **162**, 385–392.

Douglas, W. W. & Rubin, R. P. (1961). The role of calcium in the secretory response of the adrenal medulla to acetyl choline. *J. Physiol., Lond.* **159**, 40–57.

Douglas, W. W. & Rubin, R. P. (1963). The mechanism of catecholamine release from the adrenal medulla and the role of calcium in stimulus-secretion coupling. *J. Physiol., Lond.* **167**, 288–310.

Eakins, K. E. & Eakins, H. M. T. (1964). Adrenergic mechanisms and the outflow of aqueous humor from the rabbit eye. *J. Pharmac. exp. Ther.* **144**, 60–65.

Eccles, J. C. (1964). *The Physiology of Synapses.* Berlin: Springer-Verlag.

Eliasson, R., Euler, U. S. von & Stjärne, L. (1955). Studies on the release of the adrenergic neurotransmitter from the perfused ox spleen. *Acta physiol. scand.* **33**, Suppl. 118, pp. 63–69.

Euler, U. S. von, Björkman, S. & Orwén, I. (1955). Diurnal variations in the excretion of free and conjugated noradrenaline and adrenaline in urine from healthy subjects. *Acta physiol. scand.* **33**, Suppl. 118, pp. 10–16.

Euler, U. S. von, Franksson, C. & Hellström, J. (1954). Adrenaline and noradrenaline output in urine after unilateral and bilateral adrenalectomy in man. *Acta physiol. scand.* **31**, 1–8.

Euler, U. S. von & Luft, R. (1951). Noradrenaline output in urine after infusion in man. *Br. J. Pharmac. Chemother.* **6**, 286–288.

Exley, K. A. (1957). The blocking action of choline-2:6-xylyl ether bromide on adrenergic nerves. *Br. J. Pharmac. Chemother.* **12**, 297–305.

Exley, K. A. (1960). The persistence of adrenergic nerve conduction after TM 10 or bretylium in the cat. In *Adrenergic Mechanisms.* CIBA. London: Churchill.

Ferry, C. B. (1963). The sympathomimetic effect of acetylcholine on the spleen of the cat. *J. Physiol., Lond.* **167**, 487–504.

Finkleman, B. (1930). On the nature of inhibition in the intestine. *J. Physiol., Lond.* **70**, 145–157.

Fischer, J. E., Horst, W. D. & Kopin, I. J. (1965). β-Hydroxylated amines as false neurotransmitters. *Br. J. Pharmac. Chemother.* **24**, 477–484.

Folkow, B. (1952). Impulse frequency in sympathetic vasomotor fibres correlated to the release and elimination of the transmitter. *Acta physiol. scand.* **25**, 49–76.

Gaddum, J. H. & Kwiatkowski, H. (1939). Properties of the substance liberated by adrenergic nerves in the rabbit ear. *J. Physiol., Lond.* **96**, 385–391.

Gessa, G. L., Costa, E., Kuntzman, R. & Brodie, B. B. (1962). On the mechanism of norepinephrine release by α-methyl-metatyrosine. *Life Sciences* **1**, 353–360.

Gillespie, J. S. & Kirpekar, S. M. (1965). Uptake and release of H^3-noradrenaline by the splenic nerves. *J. Physiol., Lond.* **178**, 44–45*P*.

Gray, J. A. B. & Diamond, J. (1957). Pharmacological properties of sensory receptors and their relation to those of the autonomic nervous system. *Br. med. Bull.* **13**, 185–188.

Haefely, W., Hürlimann, A. & Thoenen, H. (1964). A quantitative study of the effect of cocaine on the response of the cat nictitating membrane to nerve stimulation and to injected noradrenaline. *Br. J. Pharmac. Chemother.* **22**, 5–21.

Haefely, W., Hürlimann, A. & Thoenen, H. (1965). Relation between the rate of stimulation and the quantity of noradrenaline liberated from sympathetic nerve endings in the isolated perfused spleen of the cat. *J. Physiol., Lond.* **181**, 48–58.

Haefely, W., Thoenen, H. & Hürlimann, A. (1965). The effect of sympathetic nerve stimulation in cats pretreated with α-methyl-meta-tyrosine. *Life Sciences* **4**, 913–918.

Hamberger, B., Norberg, K. A. & Sjökvist, F. (1965). Correlated studies on monoamines and acetylcholinesterase in sympathetic ganglia, illustrating the distribution of adrenergic and cholinergic neurones. In *Pharmacology of Cholinergic and Adrenergic Transmission*, ed. Koelle, Douglas and Carlsson. Oxford: Pergamon Press.

Hertting, G. (1965). Effect of drugs and sympathetic denervation on noradrenaline uptake and binding in animal tissues. In *Pharmacology of Cholinergic and Adrenergic Transmission*, ed. Koelle, Douglas and Carlsson. Oxford: Pergamon Press.

Hertting, G. & Axelrod, J. (1961). Fate of tritiated noradrenaline at the sympathetic nerve endings. *Nature, Lond*, **192**, 172–173.

Hey, P. & Willey, G. L. (1954). Choline, 2:6-xylyl ether bromide; an active quaternary local anaesthetic. *Br. J. Pharmac. Chemother.* **9**, 471–475.

Holzbauer, M. & Vogt, M. (1956). Depression by reserpine of the noradrenaline concentration in the hypothalamus of the cat. *J. Neurochem.* **1**, 8–11.

Huković, S. (1959). Isolated rabbit atria with sympathetic supply. *Br. J. Pharmac. Chemother.* **14**, 372–376.

Huković, S. & Muscholl, E. (1962). Die Noradrenalin-Abgabe aus dem isolierten Kaninchenherzen bei sympathischer Nervenreizung und ihre pharmakologische Beeinflussung. *Arch. exp. Path. Pharmak.* **244**, 81–96.

Kakimoto, Y. & Armstrong, M. D. (1962). On the identification of octopamine in mammals. *J. biol. Chem.* **237**, 422–427.

Katz, B. (1962). The transmission of impulses from nerve to muscle, and the sub-cellular unit of synaptic action. *Proc. R. Soc.* B **155**, 455–477.

Kirshner, N., Sage, H. J., Smith, W. J. & Kirshner, A. G. (1966). Release of adrenaline, noradrenaline and intragranular protein from the adrenal medulla. *Fed. Proc.* **25**, 735.

Koelle, G. B. (1965). The roles of acetylcholine and acetylcholinesterase in junctional transmission. In *Pharmacology of Cholinergic and Adrenergic Transmission*, ed. Koelle, Douglas and Carlsson. Oxford: Pergamon Press.

Kopin, I. J., Fischer, J. E., Musacchio, J. & Horst, W. D. (1964). Evidence for a false neurochemical transmitter as a mechanism for the hypotensive effect of monoamine oxidase inhibitors. *Proc. natn. Acad. Sci. U.S.A.* **52**, 716–721.

Kopin, I. J., Fischer, J. E., Musacchio, J., Horst, W. D. & Weise, V. K. (1965). 'False neurochemical transmitters' and the mechanism of sympathetic blockade by monoamine oxidase inhibitors. *J. Pharmac. exp. Ther.* **147**, 186–193.

Laverty, R., Sharman, D. F. & Vogt, M. (1965). Action of 2,4,5-trihydroxyphenylethylamine on the storage and release of noradrenaline. *Br. J. Pharmac. Chemother.* **24**, 549–560.

Liu, A. C. & Rosenblueth, A. (1935). Reflex liberation of circulating sympathin. *Am. J. Physiol.* **113**, 555–559.

Loewi, O. (1921). Über humorale Uebertragbarkeit der Herznervenwirkung. *Pflügers Arch. ges Physiol.* **189**, 239–242.

Maître, L. & Staehelin, M. (1963). Effect of α-methyl-DOPA on myocardial catecholamines. *Experientia* **19**, 573–575.

Maître, L. & Staehelin, M. (1965). Presence of α-methyl-noradrenaline ('Corbasil') in the heart of guinea pigs treated with metaraminol ('Aramine'). *Nature, Lond.* **206**, 723–724.

Mellander, S. (1960). Comparative studies on the adrenergic neuro-hormonal control of resistance and capacitance blood vessels in the cat. *Acta physiol. scand.* **50**, Suppl. 176.

Mirkin, B. L. & Bonnycastle, D. D. (1954). A pharmacological and chemical study of humoral mediators in the sympathetic nervous system. *Am. J. Physiol.* **178**, 529–534.

Musacchio, J., Fischer, J. E. & Kopin, I. J. (1965). Effect of chronic sympathetic denervation on subcellular distribution of some sympathomimetic amines. *Biochem. Pharmac.* **14**, 898–900.

Musacchio, J., Kopin, I. J. & Weise, V. K. (1965). Subcellular distribution of some sympathomimetic amines and their β-hydroxylated derivatives in the rat heart. *J. Pharmac. exp. Ther.* **148**, 22–28.

Muscholl, E. (1966). Autonomic nervous system: Newer mechanisms of adrenergic blockade. *A. Rev. Pharmacol.* **6**, 107–128.

Muscholl, E. & Maître, L. (1963). Release by sympathetic stimulation of α-methyl noradrenaline stored in the heart after administration of α-methyl dopa. *Experientia* **19**, 658–659.

Muscholl, E. & Vogt, M. (1958). Action of reserpine on the peripheral sympathetic system. *J. Physiol., Lond.* **141**, 132–155.

Peart, W. S. (1949). The nature of splenic sympathin. *J. Physiol., Lond.* **108**, 491–501.

Porter, C. C., Totaro, J. A. & Stone, C. A. (1963). The effect of 6-hydroxydopamine and some other compounds on the concentration of norepinephrine in hearts of mice. *J. Pharmac. exp. Ther.* **140**, 308–316.

Rosell, S., Kopin, I. J. & Axelrod, J. (1963). Fate of H^3-norepinephrine in skeletal muscle before and following sympathetic stimulation. *Am. J. Physiol.* **205**, 317–321.

Rosell, S., Kopin, I. J. & Axelrod, J. (1964). Release of tritiated epinephrine following sympathetic nerve stimulation. *Nature, Lond.* **201**, 301.

Rosenblueth, A. (1950). *The Transmission of Nerve Impulses at Neuroeffector Junctions and Peripheral Synapses.* New York: Wiley.

Schanker, L. S. & Morrison, A. S. (1965). Physiological disposition of guanethidine in the rat and its uptake by heart slices. *Int. J. Neuropharmac.* **4**, 27–39.

Schümann, H. J. & Philippu, A. (1963). Zum Mechanismus der durch Calcium und Magnesium verursachten Freisetzung der Nebennierenmark Hormone. *Arch. exp. Path. Pharmak.* **244**, 466–476.

Shore, P. A., Busfield, O. & Alpers, H. S. (1964). Binding and release of metaraminol: mechanism of norepinephrine depletion by α-methyl-*m*-tyrosine. *J. Pharmac exp. Ther.* **146**, 194–199.

Sjoerdsma, A., Lovenberg, W., Oates, J. A., Crout, J. R. & Udenfriend, S. (1959). Alterations in the pattern of amine excretion in man produced by monoamine oxidase inhibitor. *Science* **130**, 225.

Sjökvist, F. (1963). The correlation between the occurrence and localization of acetylcholinesterase-rich cell bodies in the stellate ganglion and the outflow of cholinergic sweat secretory fibres to the fore paw of the cat. *Acta physiol. scand.* **57**, 339–351.

Stone, C. A. & Porter, C. C. (1966). Methyl dopa and adrenergic nerve function. (Second International Catecholamine Symposium.) *Pharmac. Rev.* **18**, 569–576.

Sundin, T. (1958). The effect of body posture on the urinary excretion of adrenaline and noradrenaline. *Acta med. scand.* **161**, Suppl. 336.

Thoenen, H., Hürlimann, A. & Haefely, W. (1963). The effect of postganglionic sympathetic stimulation on the isolated, perfused spleen of the cat. Simultaneous determination of norepinephrine output and changes in volume and vascular resistance. *Helv. physiol. pharmac. Acta* **21**, 17–26.

Thoenen, H., Hürlimann, A. & Haefely, W. (1964*a*). The effect of sympathetic nerve stimulation on volume, vascular resistance and norepinephrine output in the isolated perfused spleen of the cat. *J. Pharmac. exp. Ther.* **143**, 57–63.

Thoenen, H., Hürlimann, A. & Haefely, W. (1964*b*). Wirkungen von Phenoxybenzamin, Phentolamin und Azapetin auf adrenergische Synapsen der Katzenmilz. *Helv. physiol. pharmac. Acta* **22**, 148–161.

Udenfriend, S. & Zaltzman-Nirenberg, P. (1964). On the mechanism of norepinephrine depletion by aramine. *Life Sciences* **3**, 695–702.

7

THE UPTAKE OF CATECHOLAMINES BY SYMPATHETIC NERVES

Considerable interest has centred on the role of tissue uptake mechanisms in the physiological inactivation of catecholamines. Although this concept has only received direct experimental support in recent years, various earlier findings, some of which have been described in preceding chapters, suggested that adrenaline and noradrenaline were inactivated by some non-metabolic process. The discovery of the ability of postganglionic sympathetic neurones to accumulate exogenous catecholamines and the relevance of this process in the inactivation of catecholamines will be reviewed in this chapter.

1. Discovery of the ability of peripheral tissues to accumulate exogenous catecholamines

The possibility that exogenous catecholamines might be taken up into storage sites in peripheral tissues was suggested more than thirty years ago by Burn (1932). Raab & Humphreys (1947) and Raab & Gigee (1953, 1955) were the first to demonstrate an increase in the catecholamine content of cat and dog hearts after the administration of large doses of adrenaline or noradrenaline *in vivo*. Nickerson, Berghout & Hammerström (1950) also reported large increases in the adrenaline content of the rat heart after the administration of large doses of adrenaline. Euler (1956), however, failed to demonstrate increases in the catecholamine content of various cat tissues after the administration of smaller doses of catecholamines. Early studies with C^{14}-labelled adrenaline showed that the radioactive catecholamine accumulated in the adrenal medulla and in several other tissues after intravenous injection (Schayer, 1951; Udenfriend & Wyngaarden, 1956; De Schaepdryver & Kirshner, 1961).

It was not until radioactively labelled catecholamines of high specific activity became available that similar studies could be performed using injected doses of adrenaline or noradrenaline small enough to be comparable to those likely to be encountered under physiological conditions. The first demonstration of the importance of tissue uptake in the disposition of circulating catecholamines was made by Axelrod, Weil-Malherbe & Tomchick (1959). After the intravenous administration of relatively small doses of H^3-adrenaline (0·1 mg/kg) to mice, the unchanged hormone disappeared in two phases. In the first 5 min after injection there was a rapid metabolism of approximately 70 % of the injected dose, largely by *O*-methylation. The remaining 30 %, however, disappeared only slowly thereafter, detectable amounts of unchanged adrenaline being present in animals killed several hours after the original injection. Other experiments on the fate of H^3-adrenaline after intravenous infusion in cats showed that unchanged adrenaline disappeared rapidly from the plasma while at the same time unchanged H^3-adrenaline accumulated in various peripheral tissues such as heart, spleen, lung and kidney. These results showed that after the intravenous administration of adrenaline a substantial proportion of the injected dose was inactivated by a rapid transfer from the circulation into peripheral tissues. In a subsequent study Whitby, Axelrod & Weil-Malherbe (1961) performed similar experiments with H^3-noradrenaline in cats and mice. Again it was found that tissue uptake operated to remove the intravenously administered catecholamine from the circulation. The accumulation of noradrenaline in tissues was found to be greater than that of adrenaline; in the mouse some 60 % of the injected dose of noradrenaline was inactivated by tissue uptake after the injection of a dose of 0·03 mg/kg (Fig. 7.1).

At about the same time Muscholl (1960, 1961), using sensitive bioassay and fluorimetric assay techniques, demonstrated that an appreciable accumulation of noradrenaline in the rat heart was found after the administration of small intravenous doses of noradrenaline (0·08 mg/kg). Using fluorimetric assay techniques Strömblad & Nickerson (1961) also demonstrated an accumulation of noradrenaline and adrenaline in the heart and salivary glands of rats after the administration of approximately 4 mg/kg of either hormone. These authors suggested that the uptake of catecholamines might represent an important mechanism for the physiological inactivation of catecholamines.

There have been many subsequent reports confirming the ability of various peripheral tissues to accumulate exogenous catecholamines.

Increases in the catecholamine content have been demonstrated in various organs after the administration of catecholamines *in vivo* in the cat and rat (Pennefather & Rand, 1960; Harvey & Pennefather, 1962; Bhagat, 1963; Crout, 1964). Noradrenaline uptake has also been

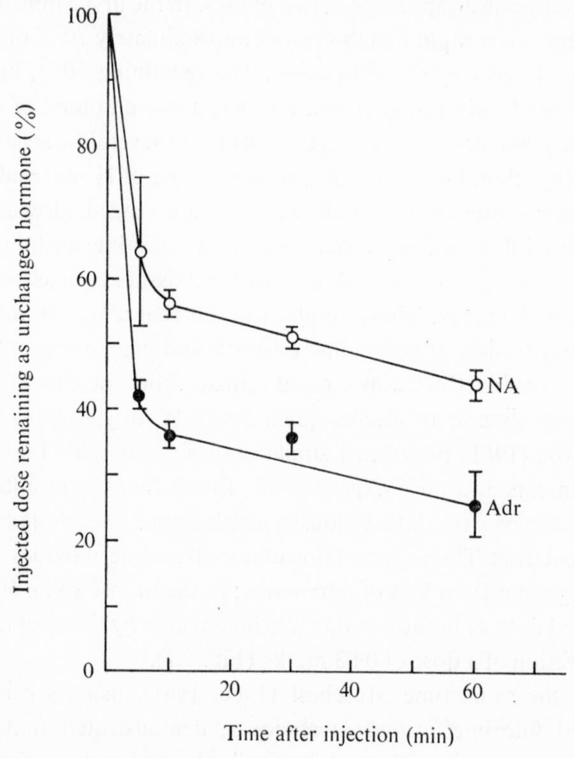

Fig. 7.1. Disappearance of noradrenaline and adrenaline in mice after the intravenous injection of 1 μg of either compound. Each point is the mean value for 3–8 animals with standard error of the mean.

demonstrated in isolated hearts or atria of the guinea-pig, rabbit, dog and rat (Muscholl, 1960, 1961; Burn & Burn, 1961; Chidsey, Kahler, Kelminson & Braunwald, 1963; Axelrod *et al.* 1962; Kopin, Hertting & Gordon, 1962; Lindmar & Muscholl, 1964; Fawaz & Simaan, 1963; Trendelenburg & Crout, 1964), and in the perfused cat spleen (Gillespie & Kirpekar, 1963, 1965*a*; Blakeley & Brown, 1963) and perfused rabbit kidney (Inouye & Tanaka, 1964). The presence of a blood–brain barrier to catecholamines has hitherto prevented the study of catecholamine uptake in brain tissue *in vivo*, although studies have been performed with brain slices incubated *in vitro* (Dengler, Spiegel & Titus, 1961; Hamberger & Masuoka, 1965). Recently, however, studies of the uptake, storage and metabolism of small doses of H^3-noradrenaline in the intact rat brain have been reported in which the labelled catecholamine was administered directly into the lateral ventricle (Glowinski, Kopin & Axelrod, 1965; Glowinski & Axelrod, 1966).

2. Evidence that catecholamine uptake occurs in sympathetic nerves

(*a*) *Distribution of accumulated noradrenaline in tissues*

In the experiments of Whitby *et al.* (1961) on the distribution of H^3-noradrenaline in various tissues of the cat after noradrenaline infusion, it was found that the uptake of noradrenaline was greatest in tissues with a rich sympathetic innervation, such as the heart. There was an approximate correlation between the amount of H^3-noradrenaline accumulated in a given tissue and the richness of the adrenergic innervation, as measured by the endogenous noradrenaline content of the tissue. In subsequent studies it has been shown that the uptake of H^3-noradrenaline into tissues after intravenous injection depends on two factors: the proportion of the cardiac output delivered to the tissue and the endogenous noradrenaline content. If allowance is made for regional differences in blood flow to various tissues, there is a good correlation between the amount of H^3-noradrenaline taken up and the endogenous noradrenaline content (Wurtman, Kopin, Horst & Fischer, 1964; Kopin, Gordon & Horst, 1965). In the guinea-pig heart and in the isolated perfused rat heart there is a significant correlation between the amount of H^3-noradrenaline taken up by the tissue and the endogenous content of noradrenaline (Crout, 1964; L. L. Iversen, unpublished observations).

(*b*) *Denervation and immunosympathectomy*

In tissues in which a normal sympathetic nerve supply is lacking, the ability to take up exogenous catecholamines is severely impaired. This has been demonstrated after surgical sympathectomy and after immunosympathectomy. After superior cervical ganglionectomy the uptake of catecholamines in tissues innervated by this ganglion is severely reduced (Hertting, Axelrod, Kopin & Whitby, 1961; Strömblad & Nickerson, 1961; Fischer, Kopin & Axelrod, 1965). The uptake of noradrenaline in the heart is also severely impaired after surgical sympathectomy; this has been achieved either by ganglionectomy (Hertting & Schiefthaler, 1964; Hertting, 1965) or by autotransplantation of the organ (Potter, Cooper, Willman & Wolfe, 1965). The uptake of H^3-noradrenaline is also markedly reduced in various tissues of immunosympathectomized rats and mice, in which the development of the sympathetic nervous system was suppressed by the administration of nerve growth factor antiserum to newborn animals (Iversen, 1965; Iversen, Glowinski & Axelrod, 1966; Zaimis, Berk & Callingham, 1965). These findings suggest that the uptake of catecholamines occurs mainly in sympathetic nerve terminals, but some caution should be used in making this interpretation. Strömblad (1959) reported that the uptake of C^{14}-adrenaline was higher in denervated cat salivary glands than in normal glands. In all denervation experiments a small uptake of noradrenaline has been found in the denervated tissues. Fischer *et al.* (1965) present evidence to suggest that this residual uptake represents an accumulation in an extraneuronal site in the tissue. Andén, Carlsson & Waldeck (1963) reported that the uptake of H^3-noradrenaline in the denervated salivary glands of the rat was not significantly lower than in the normal glands if the animals were treated with a combination of an inhibitor of catechol-*O*-methyl transferase and an inhibitor of monoamine oxidase, suggesting that under certain conditions the extraneuronal uptake may be quantitatively important. The denervation experiments, nevertheless, seem to indicate that the presence of an intact adrenergic innervation is necessary for the normal uptake and retention of exogenous catecholamines.

(*c*) *Radioautographic evidence*

After the injection of H^3-noradrenaline, the labelled catecholamine in the heart and pineal glands of the rat was shown to be localized in postganglionic sympathetic nerve terminals by a combination of radioautography and electron microscopy (Wolfe, Potter, Richardson & Axelrod, 1962; Wolfe & Potter, 1963) (Plate 6). Similar experiments

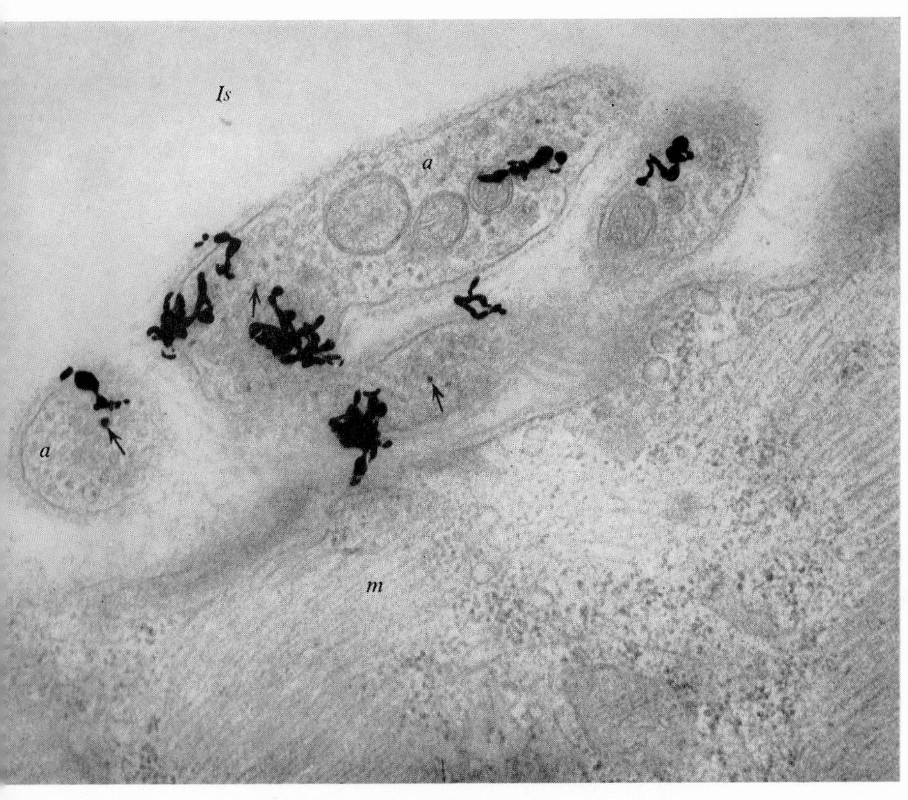

Localization of H^3-noradrenaline in adrenergic nerve terminals by means of electron microscopic ›toradiography. Rat, right atrium; 30 min after the intravenous injection of H^3-noradrenaline. ›ur terminal adrenergic axons (*a*) in an interstitial space (*Is*), close to the surface of a cardiac muscle ›l (*m*). Characteristically coiled silver grains lie over regions of axoplasm containing granulated sicles (arrows). (Cf. Plate 4.) Autoradiographic silver grains were not seen over cardiac muscle ›lls. Magnification ×34700. From D. Wolfe, unpublished.

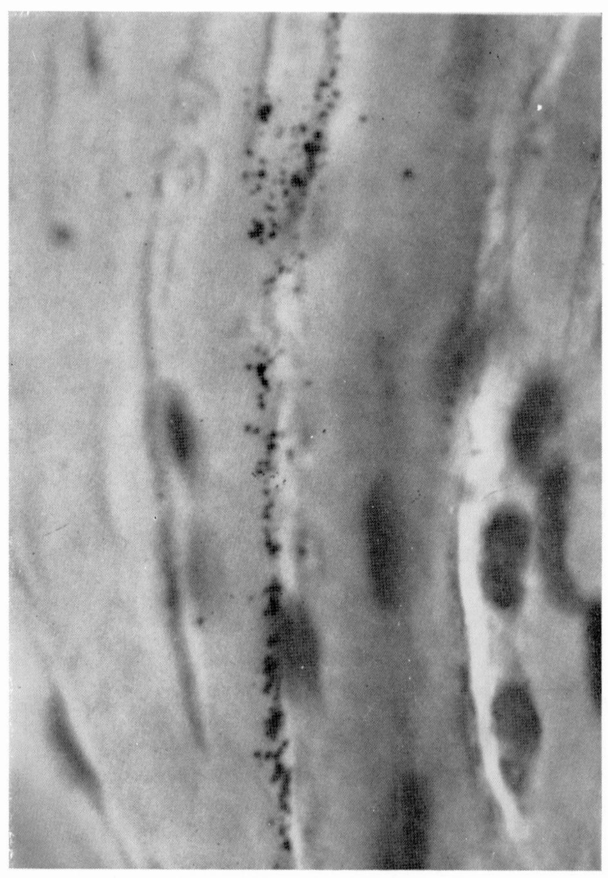

7 Distribution of H^3-noradrenaline in section of mouse heart 2h after intravenous injection of the labelled catecholamine. Radio-autograph, tissue stained in basic fuchsin. Magnification ×720. From Marks, Samorajski & Webster (1962).

using conventional autoradiographic techniques and optical microscopy have shown that H^3-noradrenaline in the brain, spleen and heart of the mouse is localized in fine nerve fibres (Samorajski & Marks, 1962; Marks, Samorajski & Webster, 1962) (Plate 7). These results provided the first direct evidence for the localization of accumulated catecholamines in nerve fibres.

(*d*) *Histochemical evidence*

Equally direct evidence for the localization of noradrenaline uptake in sympathetic nerves has been obtained in studies with the fluorescent histochemical technique for visualizing catecholamines in tissues. Hamberger, Malmfors, Norberg & Sachs (1964) showed that the administration of noradrenaline in large doses (0·5–10 mg/kg) resulted in the reappearance of fluorescent nerve fibres in the rat iris previously depleted of noradrenaline by treatment with reserpine. These experiments showed that the uptake of noradrenaline occurred not only at sympathetic nerve terminals but also in preterminal axons and in the postganglionic sympathetic nerve cell bodies in the cervical sympathetic ganglion, suggesting that the uptake of noradrenaline may occur at any point on the surface of the postganglionic sympathetic neurone. Malmfors (1965) was able to observe an accumulation of noradrenaline in the sympathetic nerve plexus of the rat iris after the injection of very small amounts of noradrenaline into the anterior chamber of the eye. Hamberger & Matsuoka (1965) demonstrated that the uptake of noradrenaline by brain slices incubated in noradrenaline *in vitro* was also localized in the fine nerve fibres in this tissue, which normally contain the endogenous catecholamine. Gillespie & Kirpekar (1965*b*) have combined the use of the fluorescent histochemical technique with radioautography in an elegant demonstration that exogenous H^3-noradrenaline infused into the cat spleen is accumulated in the nerve fibres which contain the endogenous catecholamine.

(*e*) *Other evidence*

After the administration of H^3-noradrenaline, the labelled catecholamine can be released from the cat spleen by stimulation of the splenic nerve (Hertting & Axelrod, 1961; Gillespie & Kirpekar, 1965*c*; Fischer & Iversen, 1966), indicating that it is present in the same store as the endogenous catecholamine. The rate of disappearance of labelled noradrenaline from the rat salivary gland and iris was slowed by decentralization, and the rate of disappearance from the heart was slowed by the administration of ganglionic blocking agents, indicating that the disappearance of H^3-noradrenaline from these tissues was dependent on sympathetic nerve activity (Hertting, Potter & Axelrod, 1962).

3. Properties of catecholamine uptake process

(*a*) *General*

Many studies have demonstrated that the uptake of catecholamines proceeds against a considerable concentration gradient. Axelrod *et al.* (1959) and Whitby *et al.* (1961) showed that the concentration of labelled noradrenaline and adrenaline accumulated in various tissues of the cat during infusion of the labelled catecholamines rose to levels considerably greater than those found in the plasma. Dengler *et al.* (1961) found that brain or heart slices incubated in media containing H^3-noradrenaline accumulated the labelled amine to levels up to five times those in the medium. In the intact tissue even more impressive concentration ratios between tissue and medium may be attained. In the isolated rat heart perfused with a medium containing low concentrations (10–20 ng/ml) of noradrenaline the concentration of noradrenaline accumulating in the tissue rises to levels thirty or forty times those in the perfusing medium (Iversen, 1963; Lindmar & Muscholl, 1964). If it is considered that catecholamine uptake occurs almost entirely into the sympathetic nerve plexus in the heart, it is clear that the uptake process has a remarkable ability to concentrate catecholamines. The sympathetic nerves in the rat heart can only account for a very small proportion of the total wet weight of the organ, the actual concentration ratios between exogenous noradrenaline accumulated in sympathetic nerve terminals and in the external medium must therefore be very high—perhaps exceeding 10000:1.

Sympathetic nerves have a considerable capacity to take up and retain large amounts of noradrenaline. The early experiments of Raab & Gigee (1955) demonstrated uptakes of adrenaline and noradrenaline equivalent to several times the normal content of the cat and dog heart. Subsequent studies using smaller doses of administered catecholamine have usually revealed a maximum accumulation equivalent to an approximate doubling of the endogenous noradrenaline content of tissues (Muscholl, 1961; Iversen, 1963; Crout, 1964).

(*b*) *Influence of external amine concentration*

Dengler *et al.* (1961) and Dengler, Michaelson, Spiegel & Titus (1962) found that the uptake of H^3-noradrenaline in cat heart and

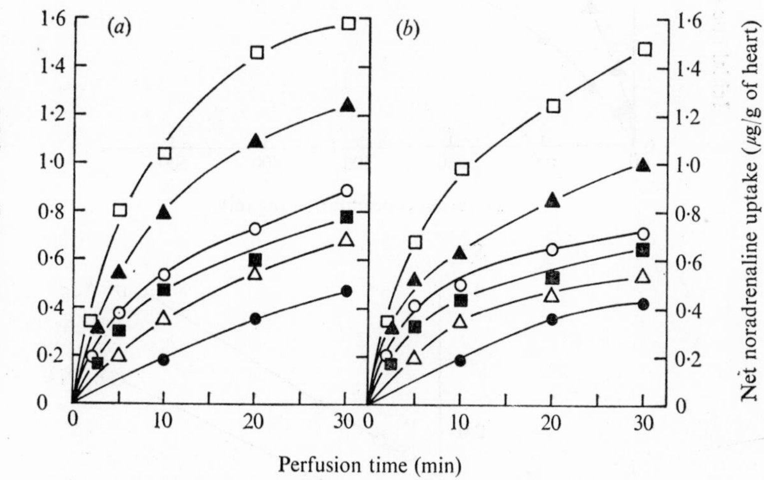

Fig. 7.2. (*a*) The uptake of H^3-noradrenaline; (*b*) the net uptake of noradrenaline by the rat isolated heart during perfusion with various concentrations of (±)-H^3-noradrenaline. ●, 10 ng/ml; △, 20 ng/ml; ■, 50 ng/ml; ○, 100 ng/ml; ▲, 500 ng/ml; □, 1 μg/ml. Each point represents the mean value of a group of six hearts. (From Iversen, 1963.)

brain slices incubated in media containing H^3-noradrenaline saturated as the concentration of noradrenaline in the medium was increased above 25 ng/ml. These authors were the first to suggest that the uptake of noradrenaline was mediated by a saturable membrane transport process. In these experiments, however, initial rates of uptake at various noradrenaline concentrations were not measured. In experiments on noradrenaline uptake in the isolated rat heart the uptake of H^3-noradrenaline was studied at various perfusion concentrations of noradrenaline in the range 10–1000 ng/ml (Fig. 7.2). From the curves of noradrenaline uptake against time thus obtained it was possible

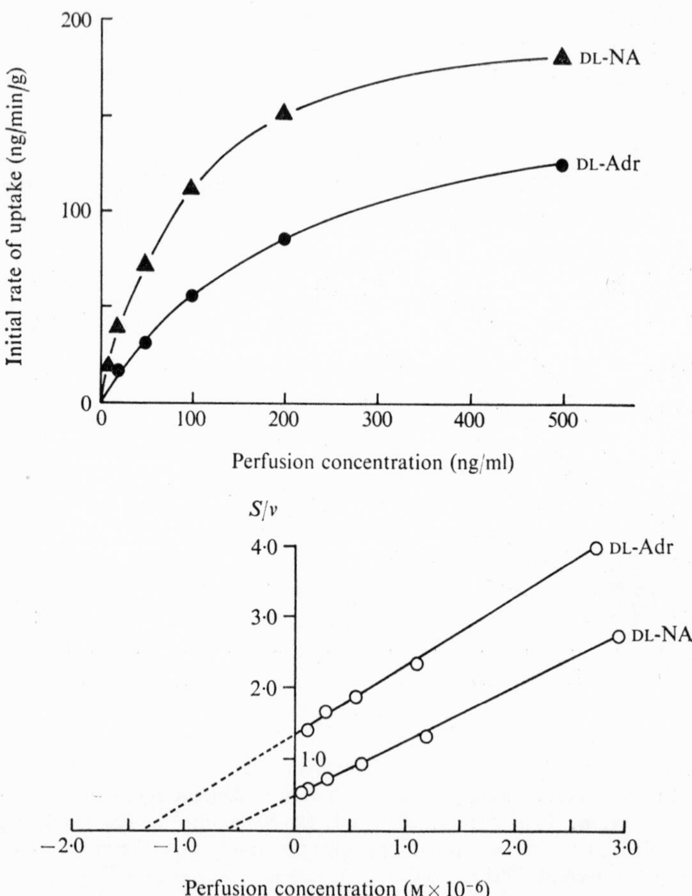

Fig. 7.3. (*a*) Initial rates of uptake of adrenaline and noradrenaline with increasing perfusion concentrations in the isolated rat heart. (*b*) Michaelis–Menten analysis of the data in (*a*), plotted as S/v against S. S, Perfusion concentration; v, initial rate of uptake. From this plot kinetic constants for noradrenaline and adrenaline uptake were determined.

to make accurate estimates of the initial rates of noradrenaline uptake at various external amine concentrations. These data were found to agree well with the classical Michaelis–Menten equation used to describe saturable enzyme/substrate interactions (Fig. 7.3). It was possible to calculate values for the maximum rates of noradrenaline

Table 7.1. *Kinetic constants for catecholamine uptake at low perfusion concentrations* ($Uptake_1$) *in the rat heart*

Compound	Michaelis constant (K_m) (M × 10^{-6})	$V_{max.}$ ± S.E. (μg/min/g heart)
DL-Noradrenaline	0·67	0·23 ± 0·005
L-Noradrenaline	0·27	0·20 ± 0·020
D-Noradrenaline	1·39	0·29 ± 0·008
DL-Adrenaline	1·40	0·19 ± 0·022

Values determined graphically (e.g. Fig. 7.5); straight lines of best fit were determined by the method of least squares and the standard error of the slope was used to determine standard errors for $V_{max.}$

uptake ($V_{max.}$) and the dissociation constant for the interaction between noradrenaline and the uptake site (K_m), these values are summarized in Table 7.1.

(c) *Stereochemical specificity*

The stereochemical specificity of the catecholamine uptake process has proved somewhat controversial. Kopin & Bridgers (1963) reported that the initial uptake of the D- and L-isomers of noradrenaline in the rat heart was the same, although the D-isomer subsequently disappeared more rapidly than the L-isomer. After the administration of D- and L-adrenaline Andén (1964) also found little difference in the accumulation of the stereoisomers in the mouse heart. Crout (1964) reported that there was an equal uptake of the stereoisomers of noradrenaline in the guinea-pig heart a short time after the injection of D- or L-noradrenaline, although it would appear that his results actually indicate a considerably greater accumulation of the L-isomer. Maickel, Beaven & Brodie (1963) and Beaven & Maickel (1964) achieved a physical separation of the stereoisomers of H^3-noradrenaline accumulated in the rat heart after the injection of DL-H^3-noradrenaline and showed that there was more than ten times more L-H^3-noradrenaline than D-H^3-noradrenaline in the tissue a few minutes after injection, indicating a marked preference for the uptake of L-noradrenaline. This conclusion was reached independently and simultaneously from our studies in the isolated rat heart (Iversen, 1963). It was found that the rate of uptake of L-noradrenaline was several times more rapid than that of D-noradrenaline, when hearts were perfused with low concentrations of the two isomers. Studies of the rate of uptake of noradrenaline during perfusion with various concentrations of L- and

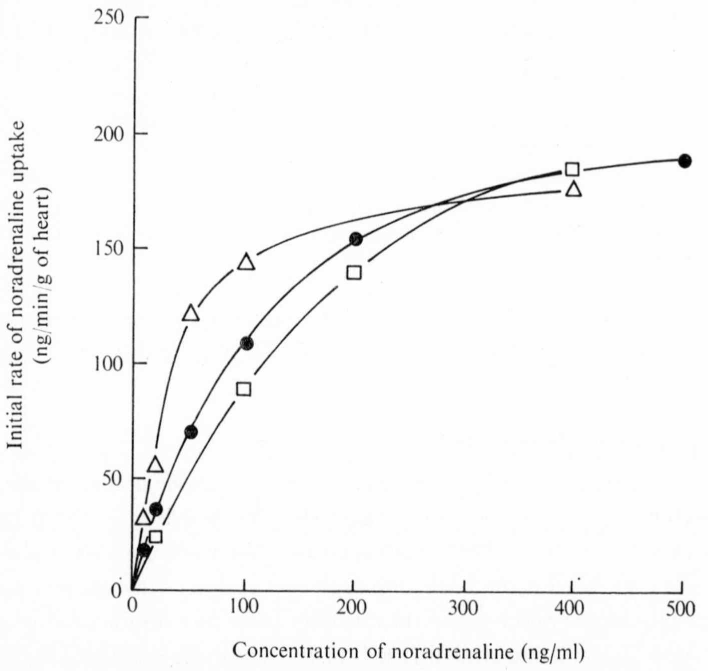

Fig. 7.4. Stereospecificity of noradrenaline uptake in the rat heart. Initial rates of uptake of D-, L- and DL-noradrenaline at various perfusion concentrations in the isolated heart. □, D-NA; ●, DL-NA; △, L-NA. (From Iversen, 1963).

D-noradrenaline showed that the uptake process had an affinity for the physiologically occurring L-isomer which was almost five times greater than the affinity for the D-isomer (Table 7.1). In membrane transport processes, however, the rate of uptake may be inversely related to the affinity of the substrate for the uptake site if high concentrations of substrate are used (Wilbrandt & Rosenberg, 1961). Thus, in the isolated rat heart the rate of uptake of D-noradrenaline was equal or greater than the rate of uptake of L-noradrenaline at high perfusion concentrations (Fig. 7.4). The demonstration of a stereochemically specific uptake of catecholamine thus depends on the use of small doses of catecholamine, and this may be one cause of the conflicting reports on this point.

(*d*) *Structural specificity*

The relative importance of tissue uptake was found to be considerably greater for noradrenaline than for adrenaline in the experiments

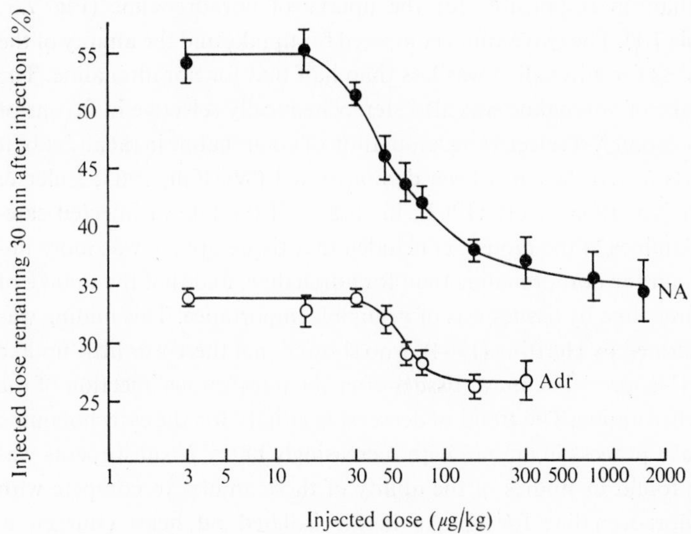

Fig. 7.5. Retention of injected catecholamines in the mouse. The quantitative importance of uptake into tissues (as measured by the amount of unchanged catecholamine retained 30 min after an intravenous injection) declines with the injection of increasing doses of H^3-noradrenaline (NA) or H^3-adrenaline (Adr). Tissue uptake is more important for noradrenaline than for adrenaline throughout the dose ranges tested. Each point is the mean value for 5–8 experiments. (Data from Iversen & Whitby, 1962.)

in which the fate of labelled catecholamines were studied in the mouse (Axelrod *et al.* 1959; Whitby *et al.* 1961). This finding was confirmed in similar studies in which it was found that the uptake of noradrenaline was quantitatively more important than that of adrenaline throughout a considerable range of injected doses (Iversen & Whitby, 1962) (Fig. 7.5). In this study we were also able to show that H^3-adrenaline and C^{14}-noradrenaline competed for uptake into common sites in mouse tissues. Strömblad & Nickerson (1961) also found that the uptake of noradrenaline into rat salivary glands was approximately twice as great as the corresponding uptake of adrenaline after the injection of equal doses of the catecholamines, and had suggested that the two amines might compete for a common storage site. In subsequent studies of the uptake of adrenaline in the isolated rat heart we were able to show that adrenaline was taken up by the saturable mechanism responsible for the uptake of noradrenaline (Fig. 7.3; Table 7.1). The two amines competed for uptake and the affinity of the process for adrenaline was less than half that for noradrenaline. The uptake of adrenaline was also stereochemically selective in favour of the L-isomer. A selective accumulation of L-adrenaline in rat and rabbit hearts *in vivo* has also been demonstrated (Westfall, 1965; Euler & Lishajko, 1965). Ross (1963), in studies of the fate of injected catecholamines in the mouse, concluded that tissue uptake was more important for noradrenaline than for adrenaline, and that the uptake of isoprenaline by tissues was of negligible importance. This finding was confirmed by Hertting (1964), who showed that there was little uptake of H^3-isoprenaline in rat tissues after the intravenous injection of the labelled amine. The trend of decreasing affinity for the catecholamine uptake process in amines with increasingly bulky *N*-substituents was also found in studies of the ability of these amines to compete with H^3-noradrenaline for uptake in the isolated rat heart (Burgen & Iversen, 1965).

In the isolated rat heart α-methyl noradrenaline is taken up by a process which is similar to that responsible for the uptake of noradrenaline (Lindmar & Muscholl, 1965; Muscholl & Weber, 1965); at equal concentrations the α-methylated catecholamine is taken up somewhat more rapidly than noradrenaline. The histochemical studies of Hamberger *et al.* (1964) and Malmfors (1965) have also demonstrated an efficient uptake of α-methyl noradrenaline into the sympathetic nerve plexus in the rat iris. Tyramine is taken up into rat salivary glands by a process which is abolished by sympathetic denervation (Carlsson & Waldeck, 1963; Fischer, Musacchio, Kopin & Axelrod, 1964), suggesting that this amine may also be accumulated by the catecholamine uptake process. α-Methyl tyramine was accumulated in the isolated rat heart by a process which could be inhibited by low concentrations of noradrenaline or by drugs which are known to inhibit the noradrenaline uptake process (Iversen, 1966). Metaraminol was avidly accumulated in rat tissues after the administration of small doses *in vivo*, this uptake was reduced by immunosympathectomy and was inhibited by drugs known to inhibit noradrenaline uptake (Shore, Busfield & Alpers, 1964). The uptake of H^3-metaraminol has also been studied by Carlsson & Waldeck (1965). It seems therefore that many structurally related amines may be taken up by the catecholamine uptake process in sympathetic nerves.

(*e*) *Summary*

Noradrenaline and other structurally related amines are taken up into postganglionic sympathetic neurones by an efficient process which is able to maintain an influx of amine against concentration gradients which may be as high as 10000:1 between tissue and external medium. This uptake process shows structural and stereochemical specificity in that noradrenaline is taken up more rapidly than its *N*-substituted derivatives and the L-isomers of noradrenaline and adrenaline are accumulated more rapidly than the D-isomers. The initial rates of noradrenaline or adrenaline uptake at various external concentrations of the two amines can be described by classical enzyme kinetics, indicating that the amines interact with a saturable uptake mechanism. It is suggested that catecholamine uptake is mediated by a membrane transport mechanism located in the axonal membrane of postganglionic sympathetic neurones.

4. Storage of accumulated catecholamines in sympathetic nerves

There is considerable evidence that once exogenous catecholamines have entered the sympathetic nerve fibre the accumulated amines undergo further redistributions into various intraneuronal storage sites.

(*a*) *Subcellular distribution studies*

Potter & Axelrod (1962, 1963*a*) examined the subcellular distribution of H^3-noradrenaline in the heart, spleen, pineal gland, vas deferens and submaxillary glands of rats after the administration of small doses of H^3-noradrenaline. After density gradient centrifugation of tissue

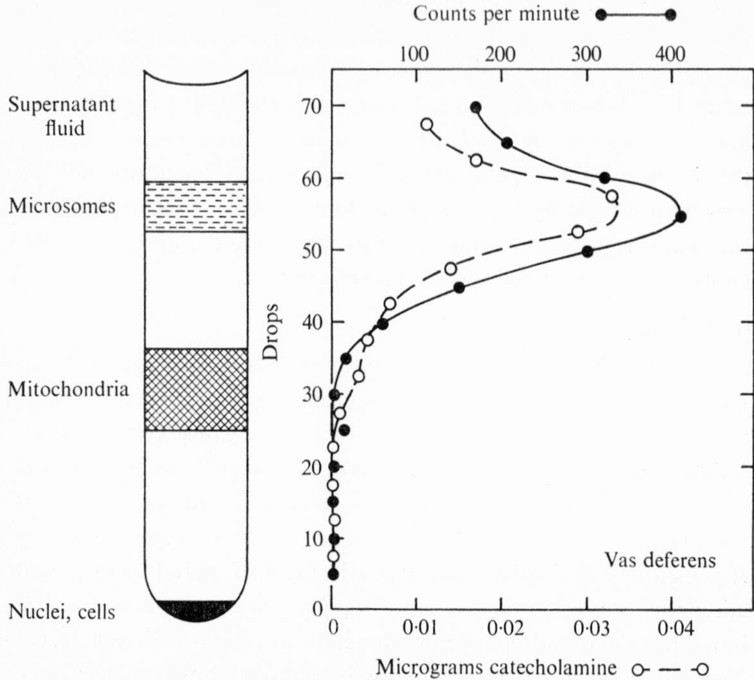

Fig. 7.6. Subcellular distribution of H^3-noradrenaline in rat vas deferens. Tissue homogenized in isotonic sucrose and layered onto a sucrose density gradient for centrifugation. Appearance of centrifuge tube at end of experiment is indicated on the left. A hole was punctured in the bottom of the tube and the contents collected in five-drop fractions. Alternate fractions were assayed for radioactivity and for endogenous noradrenaline content. The endogenous and exogenous catecholamine was distributed between the supernatant fluid and a fraction of microsomal particles lying immediately below the supernatant. (From Potter & Axelrod, 1963*a*.)

homogenates a considerable proportion of the accumulated catecholamine was found to be localized in a fraction of microsomal particles which contained the nerve storage vesicles that bind most of the endogenous noradrenaline in these tissues (Fig. 7.6). Campos & Shideman (1962) also showed that the exogenous noradrenaline accumulated in the dog heart was distributed between a microsomal particle fraction and the cell sap. These findings have been confirmed in subsequent studies (Gillis, 1964*a*; Iversen & Whitby, 1963; Michaelson, Richardson, Snyder & Titus, 1964; Snyder, Michaelson & Musacchio, 1964).

Iversen & Whitby (1963) and Stjärne (1964) found that the proportion of the total H^3-noradrenaline in rabbit or mouse heart which could be sedimented in a microsomal fraction remained constant from 3 min to up to 24 h after the injection of the labelled amine, although there were marked changes in the H^3-noradrenaline content of the tissues during this time. Accumulated catecholamines are thus rapidly distributed into intraneuronal storage particles inside the sympathetic nerve.

(*b*) *Exchange of exogenous catecholamines with endogenous*

There is considerable evidence that the noradrenaline store in adrenergic nerve terminals cannot be considered to be a single homogeneous pool; the evidence for this view is largely from pharmacological findings and will be discussed in chapter 9. There is also evidence that accumulated catecholamines do not mix freely with the endogenous noradrenaline store. In the guinea-pig heart *in vivo* or in the isolated perfused rat heart (Fig. 7.2) the H^3-noradrenaline initially accumulated after exposure to the labelled amine represents a net uptake of catecholamine which is added to the endogenous store (Crout, 1964; Iversen, 1963). The accumulated amine exchanges only slowly with the endogenous store; in the isolated rat heart the extent of this exchange did not exceed 30 % of the endogenous noradrenaline store within 1 h of perfusion. If large doses of D- or L-adrenaline are administered the accumulated adrenaline may slowly exchange with as much as 70 % of the noradrenaline stores in mouse and rat tissues (Andén, 1964; Westfall, 1965).

Chidsey & Harrison (1963) found that at short times after the administration of H^3-noradrenaline the labelled amine was preferentially released from the dog heart by nerve stimulation, the specific activity of the released noradrenaline being higher than that in the tissue as a whole. At longer times after the injection of H^3-noradrenaline (more than 6 h) the authors claimed that the H^3-noradrenaline had mixed homogeneously with the endogenous noradrenaline pool. In the cat spleen, however, 24 h after the injection of H^3-noradrenaline Fischer & Iversen (1966) found that the labelled amine was still preferentially released by nerve stimulation, the specific activity of the released amine being considerably higher than that in the tissue.

(*c*) *Disappearance of accumulated catecholamines*

Several studies have demonstrated that sympathetically innervated tissues are able to accumulate large amounts of exogenous catecholamines, amounting under certain conditions to a doubling of the normal amine stores (Muscholl, 1960; Iversen, 1963; Crout, 1964). The total noradrenaline content of the tissue, however, returns within 6–8 h to normal levels, partly by a loss of the accumulated amine and partly by an exchange of exogenous amine with the endogenous store.

After the administration of small doses of H^3-noradrenaline the labelled amine disappears from the rat heart in a complex manner. Thus, Axelrod, Hertting & Patrick (1961) found that H^3-noradrenaline disappeared from the rat heart in two main phases: the first was rapid with a half time of approximately 4–6 h, and the second was much slower, with a half time of approximately 24 h. The first phase of release was unaffected by inhibitors of either catechol-*O*-methyl transferase or monoamine oxidase; the second slower phase, however, was considerably slowed by the administration of monoamine oxidase inhibitors. The results suggested that exogenous noradrenaline entered and was released from two intracellular pools, and that the metabolic fate of the noradrenaline released from these two pools was different (Fig. 7.7). A similar conclusion was reached by Kopin & Gordon (1962, 1963*a*), who showed that after injection of H^3-noradrenaline in the rat the accumulated amine was released from the tissues in two phases, as reflected by the urinary excretion of H^3-noradrenaline and metabolites. The first rapid phase of release had a half time of about 1 h and was largely due to a release of free H^3-noradrenaline or H^3-normetanephrine, the second phase of release had a half time of about 6 h and was largely due to a release of deaminated metabolites from the tissues (Table 5.2). A study of the fate of H^3-noradrenaline in the isolated rat heart after perfusion with H^3-noradrenaline also revealed a multiphasic release of the bound amine (Kopin *et al.* 1962). The half times for the two major phases of release were 5 min and 1 h, the material released during the initial phase was largely metabolized by *O*-methylation, while that released more slowly was largely metabolized by monoamine oxidase. Montanari, Costa, Beaven & Brodie (1963) also studied the release of H^3-noradrenaline from mouse, guinea-pig and rat hearts and concluded that exogenous noradrenaline was stored in two intracellular pools with different turnover rates; half times for the two pools were reported to be 1–5 h and 24 h. Crout (1964) reported half times of approximately 4 h and 22 h for the disap-

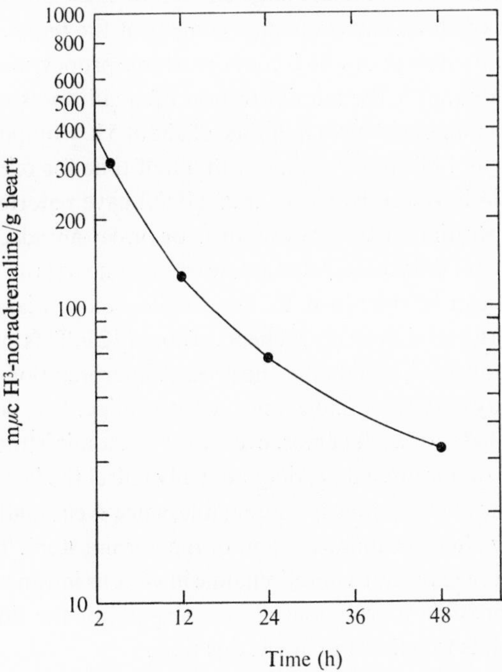

Fig. 7.7. Disappearance of H³-noradrenaline from the rat heart. The H³-noradrenaline content of the rat heart at various times after the intravenous administration of the labelled catecholamine. Each point is the mean value for six animals. (From Axelrod, Hertting & Patrick, 1961.)

pearance of H³-noradrenaline from the guinea-pig heart after the administration of small doses of H³-noradrenaline. However, if large doses of H³-noradrenaline were administered the initial rate of disappearance of the accumulated amine was much more rapid, with a half time of about 15 min. This may explain the apparently conflicting results obtained in such studies, some authors reporting a very rapid initial disappearance whereas others fail to detect this phase. The results outlined above might suggest that exogenous noradrenaline can enter and be released from at least three intraneuronal pools, which contribute the following components to the overall release curve. Component 1 represents a labile pool with a half time for release of less than 1 h, this component may only be seen after the administration of doses of exogenous noradrenaline large enough to cause an appreciable increase in the total catecholamine content of the tissue. Component 2 is a more stable pool which contains a large proportion of the amine accumulated after the administration of small doses of noradrenaline; this component has a half time of about 5 h. Component 3 is a very stable pool of noradrenaline with a half time for release of approximately 24 h. Recently, Costa *et al.* (1966) have reported that after the administration of tracer doses of L- or DL-H^3-noradrenaline the disappearance of the accumulated amine from the heart is no longer multiphasic but can be described by one simple exponential decay component, with a half time in the rat heart of about 12 h. This interesting finding, if confirmed, could alter the present interpretation of the multiphasic decay curves for exogenous noradrenaline. It is possible that the initial phases of such curves, even after the administration of very small doses of noradrenaline, does not truly reflect the behaviour of tracer amounts of exogenous noradrenaline, since even small doses may appreciably alter the amine content of the normal store. The interrelation between exogenous noradrenaline in various intraneuronal pools and the possible morphological counterparts of the different release components remains obscure at this time.

(*d*) *Biphasic entry of catecholamines into sympathetic nerves*

An analysis of the uptake of noradrenaline in the isolated rat heart against time during continuous perfusion with H^3-noradrenaline revealed the presence of at least two phases in the uptake curve (Fig. 7.8). This result suggests that noradrenaline enters at least two pools in the tissue, one is rapidly filled with a half time of less than 5 min, and a second pool fills much more slowly with a half time of approximately 20 min. The absence of the second slower phase of uptake in hearts of animals pretreated with reserpine suggested that the second phase of noradrenaline uptake might represent a slow accumulation of the amine in intraneuronal storage granules after a rapid uptake into the axoplasm of the nerve; since in the reserpinized heart the capacity of the intraneuronal granules to bind noradrenaline is abolished although the initial rate of uptake of exogenous noradrenaline in such tissues is normal (Iversen *et al.* 1965).

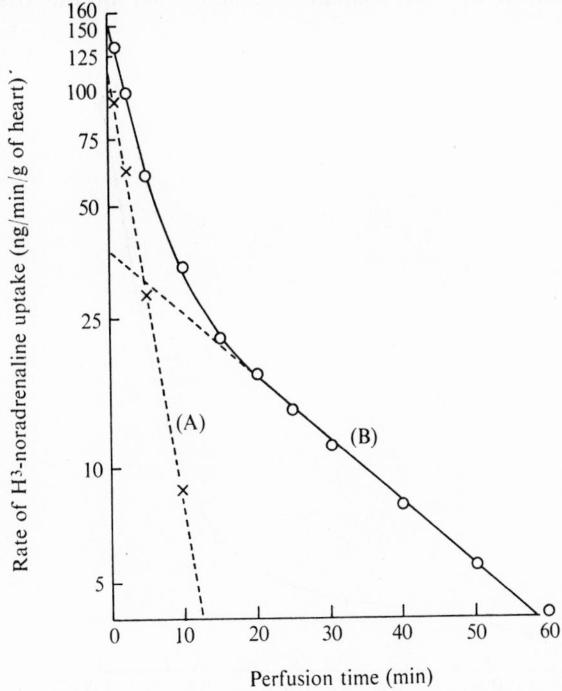

Fig. 7.8. The rate of H^3-noradrenaline uptake plotted against time during perfusion of the rat isolated heart with (±)-H^3-noradrenaline (200 ng/ml). The curve was resolved into the linear components A and B. The rate constants of these two components were 0·277 min^{-1} and 0·038 min^{-1} respectively. ○, Values derived from the uptake curve for H^3-noradrenaline at this perfusion concentration; ×, values derived by subtraction of B. (From Iversen, 1963.)

5. A second uptake process in the isolated rat heart (Uptake$_2$)

In further experiments on the uptake of catecholamines in the isolated perfused rat heart a second type of uptake was discovered (Iversen, 1965). When hearts were perfused with very high concentrations of adrenaline or noradrenaline (1–40 μg/ml) a surprisingly rapid uptake of catecholamine occurred. The accumulation of adrenaline or noradrenaline under these conditions was far too rapid to be explicable in terms of the uptake mechanism previously describ (Fig. 7.9). The second uptake process ($Uptake_2$) was, however, cle y not explicable in terms of a diffusion of catecholamines into the t ue, since the concentrations of catecholamine accumulating in the tissue

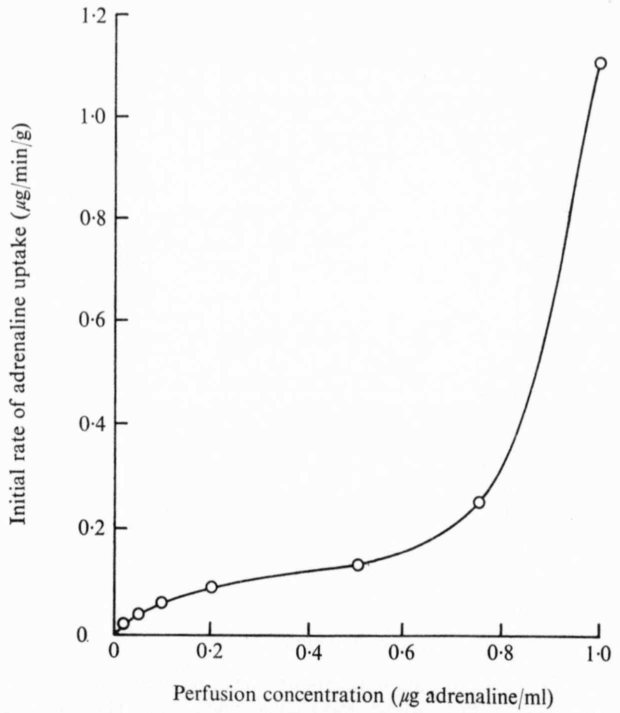

Fig. 7.9. Initial rates of adrenaline uptake in the isolated rat heart with increasing perfusion concentrations of adrenaline, showing the emergence of a second uptake mechanism ($Uptake_2$) as the perfusion concentration was raised above 0·5 µg/ml. (From Iversen, 1965.)

rapidly rose to levels several times higher than those in the perfusing medium (Fig. 7.10). The initial rates of uptake of adrenaline and noradrenaline at various perfusion concentrations could again be described by Michaelis–Menten kinetics and it was thus possible to obtain values for the kinetic constants for catecholamine uptake by the second process (Table 7.2). $Uptake_2$ differed in several respects from the process described previously ($Uptake_1$). $Uptake_2$ had a higher

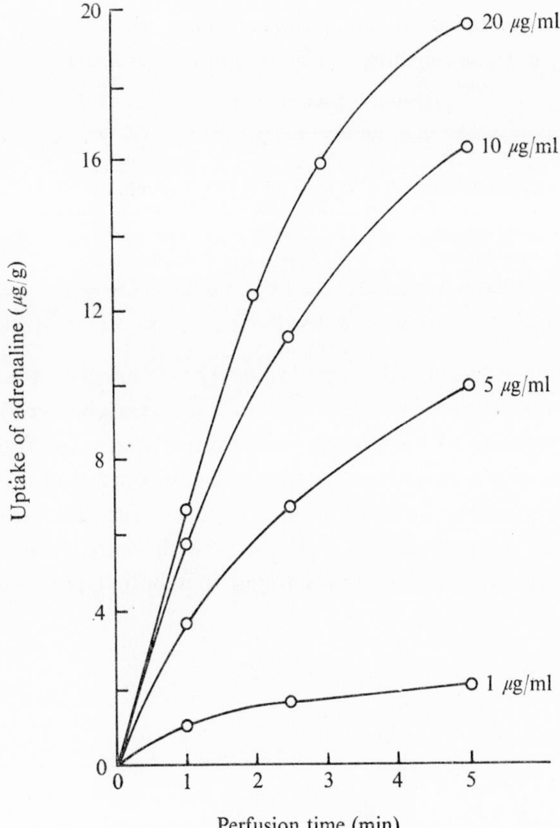

Fig. 7.10. Uptake of adrenaline by $Uptake_2$ in the isolated rat heart at various concentrations of adrenaline in the perfusion medium from 1–20 μg/ml. Values corrected for the presence of adrenaline in the extracellular space. Each value is the mean for six experiments. (From Iversen, 1965.)

affinity for adrenaline than noradrenaline, in contrast to $Uptake_1$ which favoured the accumulation of noradrenaline. In contrast to the findings with $Uptake_1$, the catecholamines accumulated by $Uptake_2$ disappeared rapidly from the tissue if perfusion was continued with a catecholamine-free medium, although a residue of approximately 2 μg/g tissue remained in the heart even after prolonged periods of wash-out (Fig. 7.11). In addition $Uptake_2$ proved to have no stereochemical specificity, the D- and L-isomers of adrenaline and nor-

Table 7.2. *Kinetic constants for catecholamine uptake at high perfusion concentrations* (*Uptake*$_2$) *in the rat heart*

(From Iversen, 1965.)

Compound	K_m (M × 10^{-6})	$V_{max.}$ ± S.E. (μg/min/g)
DL-Noradrenaline	252·0	17·0 ± 0·36
DL-Adrenaline	51·6	11·8 ± 0·25

Kinetic constants were determined as described for Uptake$_1$ (cf. Table 7.1).

adrenaline being accumulated at equal rates. There were also striking differences in the drug sensitivities of the two processes (chapter 8).

Uptake$_2$ appeared to be entirely inoperative at low perfusion concentrations and was apparently triggered into operation as the external catecholamine concentration rose above certain critical threshold levels. In preliminary experiments with hearts from immunosympathectomized rats, it was found that both Uptake$_1$ and

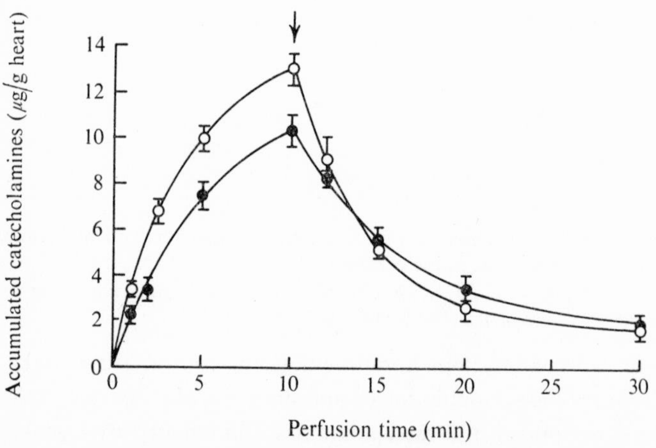

Fig. 7.11. The disappearance of catecholamines from the rat heart when perfusion for 10 min (period 0–10 min) with a medium containing (±)-adrenaline or (±)-noradrenaline at a concentration of 5 μg/ml was followed by various periods of perfusion with a catecholamine-free medium (starting at arrow). Each point is the mean value for a group of six hearts; vertical bars indicate standard errors of the means. ○, adrenaline; ●, noradrenaline. (From Iversen, 1965.)

$Uptake_2$ were severely depressed in these animals. This result suggests that both catecholamine uptake processes in the rat heart are associated with the adrenergic innervation to this organ.

The functional significance of this novel catecholamine uptake process is still unknown. It is possible that $Uptake_1$ and $Uptake_2$ are separate membrane transport processes which coexist in the terminal regions of sympathetic nerves. It is possible that $Uptake_2$, but not $Uptake_1$, is present over the whole surface of the postganglionic sympathetic neurone. In accord with this view, histochemical studies have indicated that exogenous noradrenaline accumulates in the cell bodies and preterminal axons after the administration of large doses of noradrenaline; after small doses of noradrenaline, however, an accumulation of noradrenaline was observed only in the terminal regions of the axons (Hamberger *et al.* 1964; Norberg & Hamberger, 1964).

6. Studies on catecholamine uptake in isolated storage particles

The finding that exogenous noradrenaline is rapidly accumulated in nerve storage granules in intact tissues has led to speculation about the role of these storage particles in the uptake of exogenous catecholamines. Since isolated storage granules have been shown to be able to accumulate catecholamines *in vitro*, it might seem reasonable to suppose that the uptake of noradrenaline into such granules can account for the uptake of catecholamines in intact tissues. However, in view of our present knowledge of the properties of the uptake systems in the intact tissue and in the isolated granules, this view must be regarded with some caution.

(*a*) *Uptake of catecholamines by adrenal medullary granules*

Isolated storage particles from the adrenal medulla were shown to accumulate small amounts of adrenaline or noradrenaline when incubated with high concentrations (10–20 μg/ml) of these amines (Carlsson & Hillarp, 1961). It was later found that the uptake of catecholamines was greatly increased if small amounts of adenosine triphosphate (ATP) and Mg^{2+} ions were added to the incubating medium (Kirshner, 1962; Carlsson, Hillarp & Waldeck, 1963); in the presence of these co-factors uptakes of adrenaline or noradrenaline were observed after incubation with concentrations of 1 μg/ml, but not at lower concentrations. Extensive studies of the properties of the ATP and Mg^{2+} dependent uptake of catecholamines in medullary particles have been performed (Kirshner, 1962; Carlsson *et al.* 1963; Stjärne, 1964). The uptake is limited to a maximum of some 15% of the total amine content of the particles. There is little specificity for adrenaline or noradrenaline, and several other amines such as dopamine and serotonin are also accumulated. Indeed, Carlsson *et al.* (1963) reported that the uptake of serotonin was greater than that of any of the catecholamines. The K_m for serotonin uptake in medullary particles has been estimated to be $1{\cdot}1 \times 10^{-4}$ M, while the K_m for adrenaline uptake is approximately 8×10^{-4} M (Jonasson, Rosengren & Waldeck, 1964).

The uptake of adrenaline and noradrenaline in medullary particles was reported to show some stereochemical selectivity in favour of the L-isomers (Carlsson *et al.* 1963). After entering the particles, exogenous amines do not mix freely with the endogenous catecholamines, but are nevertheless firmly bound, since incubation with non-radioactive catecholamines failed to displace accumulated C^{14}-noradrenaline. The requirement for ATP does not seem to be linked with a simultaneous incorporation of ATP into the particles, since radioactively labelled ATP did not enter the particles during the uptake of noradrenaline (Kirshner, 1962).

(b) Uptake into adrenergic nerve particles

A suspension of bovine splenic nerve particles was shown to be able to accumulate noradrenaline after the endogenous noradrenaline content had been depleted by preincubation in a catecholamine-free medium (Euler & Lishajko, 1963). Later reports from the same laboratory have suggested that two kinds of uptake process may operate in the nerve storage particles. In the absence of ATP or Mg^{2+} ions uptake occurs only at very high concentrations of external catecholamine (greater than 1 μg/ml). Under these conditions the uptake of adrenaline or noradrenaline was not stereochemically specific. In the presence of added ATP and Mg^{2+}, however, the rate of spontaneous loss of noradrenaline from particles incubated at 37 °C is greatly retarded, and under these conditions a much larger uptake of exogenous catecholamines could be demonstrated, even at concentrations below 1 μg/ml. The latter uptake was stereochemically selective in that the

L-isomers were taken up 2–3 times more rapidly than the D-isomers. When nerve particles were incubated in the presence of ATP and Mg^{2+} with radioactively labelled noradrenaline there was a considerable uptake of labelled noradrenaline, amounting to as much as 75% of the total endogenous amine content of the particles (Fig. 7.12). At the

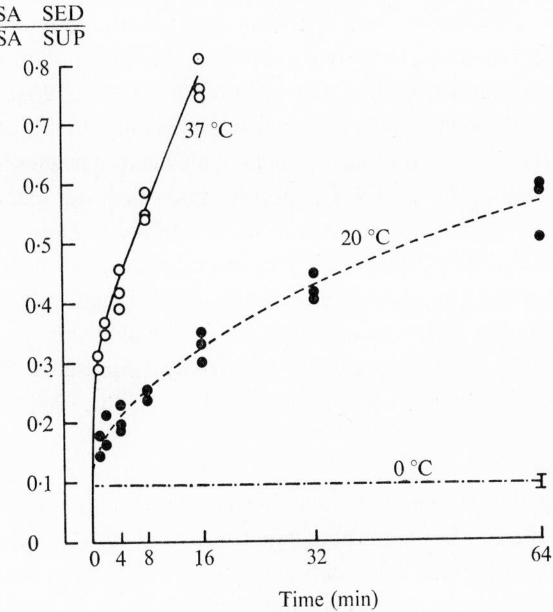

Fig. 7.12. Time course of uptake of radioactive noradrenaline in adrenergic nerve granules. Noradrenaline storage granules isolated from bovine splenic nerve were incubated with H^3-DL-noradrenaline, non-radioactive DL-noradrenaline (96 nanomoles/ml), ATP and Mg^{2+} (3 mM). (From Stjärne, 1964.)

same time little or no net increase in the total noradrenaline content was seen, suggesting that the exogenous noradrenaline had entered by exchanging with and displacing endogenous catecholamine (Euler & Lishajko, 1963, 1964; Euler, Stjärne & Lishajko, 1963; Stjärne, 1964; Stjärne & Euler, 1965).

It should be pointed out that most studies of catecholamine uptake have been performed with splenic nerve storage particles, because of the difficulty of obtaining sufficient quantities of material from other sources. The splenic nerve particles are obtained from adrenergic nerve fibres some considerable distance from the nerve terminals, and it is conceivable that such particles do not have precisely the same properties as those found at nerve terminals. More recently studies of the uptake of catecholamines in nerve storage particles isolated from the adrenergic innervation of the rat heart have been reported. The heart particles seem to have somewhat different properties from the splenic nerve particles; the heart particles were able to accumulate exogenous noradrenaline at very low concentrations of catecholamine and in the absence of added ATP or Mg^{2+} (Potter & Axelrod, 1963*b*). The uptake of noradrenaline in the heart particles was only slightly specific for L-noradrenaline. In both heart and splenic nerve particles a considerable uptake occurs at 0–4 °C, but this uptake does not seem to be due to non-specific adsorption since it can be inhibited by low concentrations of reserpine (Stjärne, 1964). Microsomal particles from rabbit heart were able to accumulate noradrenaline, dopamine and 5-hydroxytryptamine, but not DOPA or adrenaline, when incubated *in vitro*; the uptake of H^3-noradrenaline was little affected by the presence of large amounts of non-radioactive amine in the medium (Gillis, 1964*b*).

(*c*) *Conclusions*

The properties of the catecholamine uptake systems in isolated storage particles do not seem adequate to account for the uptake of catecholamines observed in intact tissues (Gillis, 1964*b*). For example, the particle uptake systems in medullary particles or in adrenergic nerve particles are not capable of any large net uptake of exogenous catecholamine, whereas, in the intact tissue a net uptake of catecholamine amounting to more than a doubling of the total amine content can be observed. The particle uptake is only seen when the particles are exposed to high concentrations of catecholamines, whilst the intact tissue is able to take up catecholamines when exposed to only very low concentrations. Furthermore, the uptake of catecholamines in medullary particles has only a low affinity for adrenaline or noradrenaline (K_m = approximately 10^{-3} M), while the uptake in the intact tissue saturates at external catecholamine concentrations of the order of 10^{-6} M or less. There are also striking differences in the sensitivities of the particle uptake systems and the intact tissue to drug inhibition. For instance, reserpine potently inhibits the uptake and release of noradrenaline in adrenergic nerve granules (Stjärne, 1964) but has no effect on the initial entry of noradrenaline into the intact tissue (see chapter 8). If the uptake of catecholamines by intraneuronal particles were solely responsible for the uptake observed in intact tissues it is also difficult to understand the rapid penetration of external catecholamines into the nerve that this would imply. The catecholamines are polar molecules which would not be expected to diffuse rapidly through lipoprotein membranes, and indeed this is evidenced by the poor absorption of orally administered catecholamines and by the lack of penetration of catecholamines through the blood–brain barrier. It seems likely that in order to account for the rapid entry of catecholamines into sympathetic nerves there must be some mechanism in the axonal membrane to facilitate the entry of the amines, and this is in accord with the conclusions of kinetic studies of the uptake of catecholamines in intact tissues, which suggest that the initial uptake of catecholamines is mediated by a membrane carrier system in the axonal membrane (Iversen, 1963). It is of course clear that, once inside the sympathetic nerve, the accumulated amines do rapidly enter the intraneuronal storage particles. This second stage in the uptake of exogenous catecholamines may serve to protect the accumulated catecholamines from degradation by intraneuronal enzymes and at the same time provide an amplification stage in the uptake allowing very high tissue/medium ratios to be attained. By removing accumulated catecholamines from the axoplasm the particle uptake may serve to reduce the apparent concentration gradient between the extracellular catecholamines and the concentration of accumulated catecholamine in the axoplasm at the inner surface of the axonal membrane.

7. Possible functions of catecholamine uptake processes

(*a*) *Termination of the actions of circulating catecholamines*

As discussed above, the uptake of catecholamines from the circulation into tissues represents a quantitatively important route for the rapid removal of catecholamines from the circulation after injection. It seems probable that an uptake of circulating catecholamines into tissues also serves as an important route for terminating the actions of catecholamines released into the circulation from the adrenal medulla. Because of the high concentrations of catechol-*O*-methyl transferase and monoamine oxidase in the liver of most species, however, circulating catecholamines can still be inactivated, though less rapidly, when the uptake of catecholamines into tissues is blocked. The evidence in favour of the view that tissue uptake is the major mechanism involved in the normal inactivation of circulating catecholamines can be briefly summarized: adrenaline and noradrenaline disappear rapidly from the circulation after an intravenous injection, and their pharmacological effects are thus rapidly terminated (Lund, 1951; Pekkarinen, 1948; Celander, 1954; Wylie, Archer & Arnold, 1960). This rapid inactivation of injected catecholamines is not markedly inhibited by inhibitors of either catechol-*O*-methyl transferase or monoamine oxidase (Crout, 1961). Inhibition of either enzyme also fails to increase the rate of urinary excretion of catecholamines (Celander & Mellander, 1955; Friend, Zileli, Hammelin & Reutter, 1958). After the injection of small doses of radioactively labelled catecholamines a large proportion of the injected dose is taken up into and remains in various peripheral tissues for long periods after the injection (Axelrod *et al.* 1959; Whitby *et al.* 1961; Iversen & Whitby, 1962).

(*b*) *Termination of the actions of noradrenaline after release from sympathetic nerves*

When sympathetic postganglionic nerves are stimulated at frequencies low enough to be comparable to those encountered physiologically, there is very little overflow of adrenergic transmitter from adrenergic neuro-effector junctions (Celander, 1954; Brown & Gillespie, 1957; Stinson, 1961; Thoenen, Hürlimann & Haefely, 1964). Noradrenaline is thus inactivated locally in the tissues in which it is released; this inactivation is not blocked when catechol-*O*-methyl transferase or monoamine oxidase or both enzymes are inhibited (Brown & Gillespie, 1957; Stinson, 1961). On the other hand the overflow of noradrenaline can be greatly increased if the uptake of noradrenaline into sympathetic nerves is inhibited (Brown, 1965). Furthermore, inhibition of noradrenaline uptake markedly potentiates the effects of noradrenaline released by nerve stimulation or administered exogenously to smooth muscle effectors, suggesting that the uptake normally competes effectively with adrenegic receptors for the noradrenaline present in the extracellular space in the vicinity of adrenergic terminals.

Koelle (1959), reviewing the evidence then available concerning the inactivation of noradrenaline, was led to the conclusion that the non-metabolic inactivation of catecholamines by 'tissue redistribution' was probably more important than metabolic degradation. Burn & Rand (1958) had also suggested that noradrenaline storage sites existed in peripheral tissues which could take up exogenous catecholamines to restore tyramine responses in reserpinized animals. The discovery of an efficient, specific noradrenaline uptake process in sympathetic nerve terminals has led to the current view that it is this process which is primarily responsible for the termination of the actions of liberated transmitter at adrenergic nerve terminals. The relatively minor role played by metabolism is illustrated by the experiments on noradrenaline uptake in the isolated perfused rat heart. When this organ is perfused with low concentrations of noradrenaline a considerable proportion of the noradrenaline content of the medium (15–30 %) is removed by uptake into the tissue; of this amount which is taken up into the tissue more than 90 % remains as unchanged noradrenaline, indicating that metabolism plays only a minor role in the inactivation of extracellular catecholamines in this organ (Kopin *et al.* 1962; Lindmar & Muscholl, 1964; Iversen *et al.* 1965). The experiments of Brown (1965) have also indicated that more than 90 % of the noradrenaline released from nerve terminals in the cat spleen is inactivated by tissue binding.

There remains some doubt, however, about the exact proportion of the released noradrenaline which is inactivated by re-uptake into sympathetic nerves. A certain proportion of the released transmitter will not be recaptured and will either be metabolized in the tissue by *O*-methylation or will overflow into the circulation and be metabolized in the liver or kidneys. Recent studies have shown that in the cat spleen the efficiency of the re-uptake process depends greatly on the rate of perfusion of the organ. Thus if the organ is perfused at a rapid rate, released noradrenaline may be washed away from the tissue before being taken up into nerve endings (Hertting & Schiefthaler, 1963; Hertting, 1965; Paton & Gillis, 1965; Fischer & Iversen, 1966) (Fig. 7.13). Thus the efficiency of re-uptake as a method for the inactivation of noradrenaline may depend considerably on the rate of blood flow through the tissue. Since, however, the released noradrenaline will tend to cause a local vasoconstriction it may be supposed that under most conditions inactivation by re-uptake will be favoured (Rosell, Kopin & Axelrod, 1963). In view of the importance of this process,

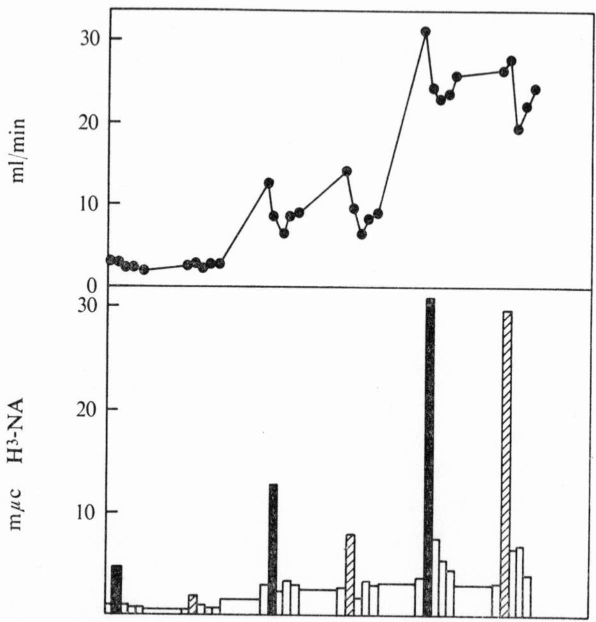

Fig. 7.13. Effect of flow rate on the H^3-noradrenaline released from the spleen by sympathetic nerve stimulation. The spleen of a cat which had received 200 μc/kg DL-H^3-noradrenaline was removed and perfused with Krebs–Ringer-bicarbonate 6 h after intravenous administration of the catecholamine. The perfusion rate was varied by alteration of the perfusion pressure. The splenic nerve was stimulated 300 times using 5 V, 2 msec impulses at a frequency of 30/sec (black) or 10/sec (striped). The outflow of H^3-noradrenaline and the perfusion rate during 1 min intervals are shown. (From Hertting, 1965.)

the inactivation of noradrenaline after neural release seems likely to be the most important role of the catecholamine uptake systems in sympathetic nerves.

(c) *The origin of noradrenaline in peripheral tissues*

Since peripheral sympathetically innervated tissues are able to accumulate circulating catecholamines, it is possible that this process might serve to maintain a constant level of noradrenaline in these tissues at the expense of catecholamines liberated into the circulation from the adrenal medulla. There is little evidence, however, that this represents an important source of noradrenaline in most peripheral tissues; as discussed in chapter 5, sympathetic nerve terminals are able to synthesize noradrenaline sufficiently rapidly to maintain a constant level of noradrenaline under most conditions. Kopin & Gordon (1963*b*) and Kopin *et al.* (1965) measured the specific activity of noradrenaline in the heart and other organs of the rat after long periods of continuous infusion of C^{14}-noradrenaline. In the rat heart the specific activity never exceeded 20 % of that of noradrenaline in the circulation, suggesting that no more than 20 % of the noradrenaline in the heart could normally derive from an uptake from the circulation. In other tissues even smaller proportions of the total content of noradrenaline could derive from this source. In demedullated animals the noradrenaline content of peripheral tissues is unaffected (Strömblad & Nickerson, 1961; Bhagat, 1963; Bhagat & Shideman, 1964). In demedullated rats, however, the latter authors reported that the rate of repletion of cardiac noradrenaline stores after reserpine was slower than in normal animals, suggesting that the uptake of circulating catecholamines may be important in replenishing peripheral noradrenaline stores under certain conditions.

While there is little evidence that an uptake of noradrenaline from the circulation is important for maintaining the noradrenaline stores in peripheral tissues it is possible that an uptake of circulating adrenaline may account for the small and variable amounts of adrenaline found in some peripheral tissues. An uptake of circulating adrenaline could account for the considerable increases in the adrenaline content of peripheral tissues when large amounts of adrenaline are released from the adrenal medulla by splanchnic nerve stimulation or by insulin administration (Raab & Gigee, 1953) or by triethyltin (Moore & Brody, 1961). Strömblad (1961) also observed a large increase in the adrenaline content of the rat heart and other tissues during insulin treatment, and this increase did not occur in demedullated animals.

REFERENCES

Andén, N. E. (1964). Uptake and release of dextro- and laevo-adrenaline in noradrenaline stores. *Acta pharmac. tox.* **21**, 59–75.

Andén, N. E., Carlsson, A. & Waldeck, B. (1963). Reserpine-resistant uptake mechanisms of noradrenaline in tissues. *Life Sciences* **2**, 889–894.

Axelrod, J., Gordon, E., Hertting, G., Kopin, I. J. & Potter, L. T. (1962). On the mechanism of tachyphylaxis to tyramine in the isolated rat heart. *Br. J. Pharmac. Chemother.* **19**, 56–63.

Axelrod, J., Hertting, G. & Patrick, R. W. (1961). Inhibition of H^3-norepinephrine release by monoamine oxidase inhibitors. *J. Pharmac. exp. Ther.* **134**, 325–328.

Axelrod, J., Weil-Malherbe, H. & Tomchick, R. (1959). The physiological disposition of H^3-epinephrine and its metabolite metanephrine. *J. Pharmac. exp. Ther.* **127**, 251–256.

Beaven, M. A. & Maickel, R. P. (1964). Stereoselectivity of norepinephrine storage sites in the heart. *Biochem. biophys. Res. Commun.* **14**, 509–513.

Bhagat, B. (1963). Effect of noradrenaline injection on the catecholamine content of the rat heart. *Archs. int. Pharmacodyn. Thér.* **146**, 47–55.

Bhagat, B. & Shideman, F. E. (1964). Repletion of cardiac catecholamines in the rat: importance of adrenal medulla and synthesis from precursors. *J. Pharmac. exp. Ther.* **143**, 77–81.

Blakeley, A. G. H. & Brown, G. L. (1963). Uptake of noradrenaline by the isolated perfused spleen. *J. Physiol., Lond.* **169**, 98–99*P*.

Brown, G. L. (1965). The Croonian Lecture 1964. The release and fate of the transmitter liberated by adrenergic nerves. *Proc. R. Soc.* B **162**, 1–19.

Brown, G. L. & Gillespie, J. S. (1957). The output of sympathetic transmitter from the spleen of the cat. *J. Physiol., Lond.* **138**, 81–102.

Burgen, A. S. V. & Iversen, L. L. (1965). The inhibition of noradrenaline uptake by sympathomimetic amines in the rat isolated heart. *Br. J. Pharmac. Chemother.* **25**, 34–49.

Burn, G. P. & Burn, J. H. (1961). Uptake of labelled noradrenaline by isolated atria. *Br. J. Pharmac. Chemother.* **16**, 344–351.

Burn, J. H. (1932). The action of tyramine and ephedrine. *J. Pharmac. exp. Ther.* **46**, 75–95.

Burn, J. H. & Rand, M. J. (1958). The action of sympathomimetic amines in animals treated with reserpine. *J. Physiol., Lond.* **144**, 314–346.

Campos, H. A. & Shideman, F. E. (1962). Subcellular distribution of catecholamines in the dog heart—effect of reserpine and norepinephrine administration. *Int. J. Neuropharmac.* **1**, 13–22.

Carlsson, A. & Hillarp, N.-Å. (1961). Uptake of phenyl- and indole-alkylamines by the storage granules of the adrenal medulla *in vitro*. *Med. exp.* **5**, 122–124.

Carlsson, A., Hillarp, N.-Å. & Waldeck, B. (1963). Analysis of the Mg^{++}-ATP dependent storage mechanisms in the amine granules of the adrenal medulla. *Acta physiol. scand.* **59**, Suppl. 215.

Carlsson, A. & Waldeck, B. (1963). *β*-Hydroxylation of tyramine *in vivo*. *Acta pharmac. tox.* **20**, 371–375.

Carlsson, A. & Waldeck, B. (1965). Mechanism of amine transport in the cell membranes of the adrenergic nerves. *Acta pharmac. tox.* **22**, 293–300.

Celander, O. (1954). The range of control exercised by the sympathico-adrenal system. *Acta physiol. scand.* **32**, Suppl. 116.

Celander, O. & Mellander, S. (1955). Elimination of adrenaline and noradrenaline from the circulating blood. *Nature, Lond.* **176**, 973–974.

Chidsey, C. A. & Harrison, D. C. (1963). Studies on the distribution of exogenous norepinephrine in the sympathetic neurotransmitter stores. *J. Pharmac. exp. Ther.* **140**, 217–223.

Chidsey, C. A., Kahler, R. L., Kelminson, L. & Braunwald, E. (1963). Uptake and metabolism of tritiated norepinephrine in the isolated canine heart. *Circulation Res.* **12**, 220–227.

Costa, E., Boullin, D. J., Hammer, W., Vogel, E. & Brodie, B. B. (1966). Interactions of drugs with adrenergic neurones. (Second International Catecholamine Symposium.) *Pharmacol. Rev.* **18**, 577–598.

Crout, J. R. (1961). Effect of inhibiting both catechol-*O*-methyl transferase and monoamine oxidase on the cardiovascular responses to norepinephrine. *Proc. Soc. exp. Biol. Med.* **108**, 482–484.

Crout, J. R. (1964). The uptake and release of H^3-norepinephrine by the guinea pig heart *in vivo*. *Arch. exp. Path. Pharmak.* **248**, 85–98.

De Schaepdryver, A. F. & Kirshner, N. (1961). Metabolism of DL-adrenaline-2-C^{14} in the cat. II. Tissue metabolism. *Archs. int. Pharmacodyn. Thér.* **131**, 433–449.

Dengler, H. J., Michaelson, I. A., Spiegel, H. E. & Titus, E. (1962). The uptake of labelled norepinephrine by isolated brain and other tissues of the cat. *Int. J. Neuropharmac.* **1**, 23–38.

Dengler, H. J., Spiegel, H. E. & Titus, E. O. (1961). Uptake of tritium labelled norepinephrine in brain and other tissues of cat *in vitro*. *Science, N.Y.* **133**, 1072–1073.

Euler, U. S. von. (1956). The catechol amine content of various organs of the cat after injections and infusions of adrenaline and noradrenaline. *Circulation Res.* **4**, 647–652.

Euler, U. S. von & Lishajko, F. (1963). Catecholamine release and uptake in isolated adrenergic nerve granules. *Acta physiol. scand.* **57**, 468–480.

Euler, U. S. von & Lishajko, F. (1964). Uptake of L- and D-isomers of catecholamines in adrenergic nerve granules. *Acta physiol. scand.* **60**, 217–222.

Euler, U. S. von & Lishajko, F. (1965). Uptake of catecholamines in the rabbit heart after depletion with decaborane. *Life Sciences* **4**, 969–972.

Euler, U. S. von, Stjärne, L. & Lishajko, F. (1963). Uptake of radioactively labelled DL-catecholamines in isolated adrenergic nerve granules with and without reserpine. *Life Sciences* **2**, 878–885.

Fawaz, G. & Simaan, J. (1963). Cardiac noradrenaline stores. *Br. J. Pharmac. Chemother.* **20**, 569–578.

Fischer, J. E. & Iversen, L. L. (1966). Effect of various factors on the output of noradrenaline from the perfused cat spleen on stimulation of the splenic nerve. (In preparation.)

Fischer, J. E., Kopin, I. J. & Axelrod, J. (1965). Evidence for extraneuronal binding of norepinephrine. *J. Pharmac. exp. Ther.* 181–185.

Fischer, J. E., Musacchio, J., Kopin, I. J. & Axelrod, J. (1964). Effects of denervation on the uptake and β-hydroxylation of tyramine in the rat salivary gland. *Life Sciences* **3**, 413–419.

Friend, G., Zileli, M. S., Hammelin, J. R. & Reutter, F. W. (1958). The effect of iproniazid on the inactivation of norepinephrine in the human. *J. clin. exp. Psychopath.* **19**, 61–68.

Gillespie, J. S. & Kirpekar, S. M. (1963). Removal of noradrenaline from the blood flowing through the spleen. *J. Physiol., Lond.* **169**, 100–101 *P*.

Gillespie, J. S. & Kirpekar, S. M. (1965 *a*). The inactivation of infused noradrenaline by the cat spleen. *J. Physiol., Lond.* **176**, 205–227.

Gillespie, J. S. & Kirpekar, S. M. (1965 *b*). The localization of endogenous and infused noradrenaline in the spleen. *J. Physiol., Lond.* **178**, 46–47 *P*.

Gillespie, J. S. & Kirpekar, S. M. (1965 *c*). Uptake and release of H^3-noradrenaline by the splenic nerves. *J. Physiol., Lond.* **178**, 44–45 *P*.

Gillis, C. N. (1964 *a*). The retention of exogenous norepinephrine by rabbit tissues. *Biochem. Pharmac.* **13**, 1–12.

Gillis, C. N. (1964 *b*). Characteristics of norepinephrine retention by a subcellular fraction of rabbit heart. *J. Pharmac. exp. Ther.* **146**, 54–60.

Glowinski, J. & Axelrod, J. (1966). Effects of drugs on the disposition of H^3-norepinephrine in the rat brain. (Second International Catecholamine Symposium.) *Pharmacol. Rev.* **18**, 775–786.

Glowinski, J., Kopin, I. J. & Axelrod, J. (1965). Metabolism of H^3-norepinephrine in the rat brain. *J. Neurochem.* **12**, 25–30.

Hamberger, B., Malmfors, T., Norberg, K. A. & Sachs, C. (1964). Uptake and accumulation of catecholamines in peripheral adrenergic neurons of reserpinized animals, studied with a histochemical method. *Biochem. Pharmac.* **13**, 841–844.

Hamberger, B. & Masuoka, D. (1965). Localization of catecholamine uptake in rat brain slices. *Acta pharmac. tox.* **22**, 363–368.

Harvey, J. A. & Pennefather, J. N. (1962). Effect of adrenaline infusions on the catechol amine content of cat and rat tissues. *Br. J. Pharmac. Chemother.* **18**, 183–189.

Hertting, G. (1964). The fate of H^3-iso-proterenol in the rat. *Biochem. Pharmac.* **13**, 1119-1128.

Hertting, G. (1965). Effects of drugs and sympathetic denervation on noradrenaline uptake and binding in animal tissues. In *Pharmacology of Cholinergic and Adrenergic Transmission* (ed. Koelle, Douglas and Carlsson), pp. 277–288. Oxford: Pergamon Press.

Hertting, G. & Axelrod, J. (1961). Fate of tritiated noradrenaline at the sympathetic nerve endings. *Nature, Lond.* **192**, 172–173.

Hertting, G., Axelrod, J., Kopin, I. J. & Whitby, L. G. (1961). Lack of uptake of catecholamines after chronic denervation of sympathetic nerves. *Nature, Lond.* **189**, 66.

Hertting, G., Potter, L. T. & Axelrod, J. (1962). Effect of decentralization and ganglionic blocking agents on the spontaneous release of H^3-norepinephrine. *J. Pharmac. exp. Ther.* **136**, 289–292.

Hertting, G. & Schiefthaler, T. (1963). Beziehung zwischen Durchflussgrosse und Noradrenalin Freisetzung bei Nervenreizung der isoliert durchströmten Katzenmilz. *Arch. exp. Path. Pharmak.* **246**, 13–14.

Hertting, G. & Schiefthaler, T. (1964). The effect of stellate ganglion excision on the catecholamine content and the uptake of H^3-norepinephrine in the heart of the cat. *Int. J. Neuropharmac.* **3**, 65–69.

Inouye, A. & Tanaka, I. (1964). Effect of tyramine, reserpine and cocaine on the noradrenaline release and uptake in the perfused rabbit kidney. *Acta physiol. scand.* **62**, 359–363.

Iversen, L. L. (1963). The uptake of noradrenaline by the isolated perfused rat heart. *Br. J. Pharmac. Chemother.* **21**, 523–537.

Iversen, L. L. (1965). The uptake of catecholamines at high perfusion concentrations in the rat isolated heart: a novel catecholamine uptake process. *Br. J. Pharmac. Chemother.* **25**, 18–33.

Iversen, L. L. (1966). Accumulation of α-methyl tyramine by the noradrenaline uptake process in the isolated rat heart. *J. Pharm. Pharmac.* **18**, 481–484.

Iversen, L. L., Glowinski, J. & Axelrod, J. (1965). The uptake and storage of H^3-norepinephrine in the reserpine-pretreated rat heart. *J. Pharmac. exp. Ther.* **150**, 173–183.

Iversen, L. L., Glowinski, J. & Axelrod, J. (1966). The physiological disposition and metabolism of norepinephrine in immunosympathectomized animals. *J. Pharmac. exp. Ther.* **151**, 273–284.

Iversen, L. L. & Whitby, L. G. (1962). Retention of injected catechol amines by the mouse. *Br. J. Pharmac. Chemother.* **19**, 355–364.

Iversen, L. L. & Whitby, L. G. (1963). The subcellular distribution of catecholamines in normal and tyramine-depleted mouse hearts. *Biochem. Pharmac.* **12**, 582–584.

Jonasson, J., Rosengren, E. & Waldeck, B. (1964). On the effects of some pharmacologically active amines on the uptake of arylalkylamines by adrenal medullary granules. *Acta physiol. scand.* **60**, 136–140.

Kirshner, N. (1962). Uptake of catecholamines by a particulate fraction of the adrenal medulla. *J. biol. Chem.* **237**, 2311–2317.

Koelle, G. B. (1959). Possible mechanisms for the termination of the physiological actions of catecholamines. *Pharmac. Rev.* **11**, 381–386.

Kopin, I. J. & Bridgers, W. (1963). Differences in D- and L-norepinephrine-H^3. *Life Sciences* **2**, 356–362.

Kopin, I. J. & Gordon, E. K. (1962). Metabolism of norepinephrine-H^3 released by tyramine and reserpine. *J. Pharmac. exp. Ther.* **138**, 351–359.

Kopin, I. J. & Gordon, E. K. (1963*a*). Metabolism of administered and drug-released norepinephrine-7-H^3 in the rat. *J. Pharmac. exp. Ther.* **140**, 207–216.

Kopin, I. J. & Gordon, E. K. (1963*b*). Origin of norepinephrine in the heart. *Nature, Lond.* **199**, 1289.

Kopin, I. J., Gordon, E. K. & Horst, W. D. (1965). Studies of uptake of L-norepinephrine-C^{14}. *Biochem. Pharmac.* **14**, 753–760.

Kopin, I. J., Hertting, G. & Gordon, E. K. (1962). Fate of norepinephrine-H^3 in the isolated perfused rat heart. *J. Pharmac. exp. Ther.* **138**, 34–40.

Lindmar, R. & Muscholl, E. (1964). Die Wirkung von Pharmaka auf die Elimination von Noradrenalin aus der Perfusionsflüssigkeit und die Noradrenalin-Aufnahme in das isolierte Herz. *Arch. exp. Path. Pharmak.* **247**, 469–492.

Lindmar, R. & Muscholl, E. (1965). Die Aufnahme von α-Methylnoradrenalin in das isolierte Kaninchenherz und seine Freisetzung durch Reserpin und Guanethidin *in vivo*. *Arch. exp. Path. Pharmak.* **249**, 529–548.

Lund, A. (1951). Elimination of adrenaline and noradrenaline from the organism. *Acta pharmac. tox.* **7**, 297–308.

Maickel, R. P., Beaven, M. A. & Brodie, B. B. (1963). Implications of uptake and storage of norepinephrine by sympathetic nerve endings. *Life Sciences* **2**, 953–958.

Malmfors, T. (1965). Studies on adrenergic nerves. *Acta physiol. scand.* **64**, Suppl. 248.

Marks, B. H., Samorajski, T. & Webster, E. J. (1962). Radioautographic localization of norepinephrine-H^3 in the tissues of mice. *J. Pharmac. exp. Ther.* **138**, 376–381.

Michaelson, I. A., Richardson, K. L., Snyder, S. H. & Titus, E. O. (1964). The separation of catecholamine storage vesicles from rat heart. *Life Sciences* **3**, 971–978.

Montanari, R., Costa, E., Beaven, M. A. & Brodie, B. B. (1963). Turnover rates of norepinephrine in hearts of intact mice, rats, and guinea-pigs using tritiated norepinephrine. *Life Sciences* **2**, 232–240.

Moore, K. E. & Brody, T. M. (1961). The effect of triethyltin on tissue amines. *J. Pharmac. exp. Ther.* **132**, 6–12.

Muscholl, E. (1960). Die Hemmung der Noradrenalin-Aufnahme des Herzens durch Reserpin und die Wirkung von Tyramin. *Arch. exp. Path. Pharmak.* **240**, 234–241.

Muscholl, E. (1961). Effect of cocaine and related drugs on the uptake of noradrenaline by heart and spleen. *Br. J. Pharmac. Chemother.* **16**, 352–359.

Muscholl, E. & Weber, E. (1965). Die Hemmung der Aufnahme von α-Methylnoradrenalin in das Herz durch sympathomimetische Amine. *Arch. exp. Path. Pharmak.* **252**, 134–143.

Nickerson, M., Berghout, J. & Hammerström, R. N. (1950). Mechanism of the acute lethal effect of epinephrine in rats. *Am. J. Physiol.* **160**, 479–484.

Norberg, K.-A. & Hamberger, B. (1964). The sympathetic adrenergic neurone. *Acta physiol. scand.* **63**, Suppl. 238.

Paton, D. H. & Gillis, C. N. (1965). Effect of altered perfusion rates on the retention of noradrenaline by the spleen. *Nature, Lond.* **208**, 391–392.

Pekkarinen, A. (1948). Studies on the chemical determination, occurrence and metabolism of adrenaline in the animal organism. *Acta physiol. scand.* **16**, Suppl. 54.

Pennefather, J. N. & Rand, M. J. (1960). Increase in noradrenaline content of tissues after infusion of noradrenaline, dopamine and L-DOPA. *J. Physiol., Lond.* **154**, 277–287.

Potter, L. T. & Axelrod, J. (1962). Intracellular localization of catecholamines in tissues of the rat. *Nature, Lond.* **194**, 581–582.

Potter, L. T. & Axelrod, J. (1963*a*). Subcellular localization of catecholamines in the tissues of the rat. *J. Pharmac. exp. Ther.* **142**, 291–298.

Potter, L. T. & Axelrod, J. (1963*b*). Properties of norepinephrine storage particles of the rat heart. *J. Pharmac. exp. Ther.* **142**, 299–305.

Potter, L. T., Cooper, T., Willman, V. L. & Wolfe, D. E. (1965). Synthesis, binding, release and metabolism of norepinephrine in normal and transplanted dog hearts. *Circulation Res.* **16**, 468–481.

Raab, W. & Gigee, W. (1953). Die Katecholamine des Herzens. *Arch. exp. Path. Pharmak.* **219**, 248–262.

Raab, W. & Gigee, W. (1955). Specific avidity of heart muscle to absorb and store epinephrine and norepinephrine. *Circulation Res.* **3**, 553–558.

Raab, W. & Humphreys, R. J. (1947). Drug action upon myocardial epinephrine–sympathin concentration and heart rate (nitroglycerine, papaverine, priscol, dibenamine-hydrochloride). *J. Pharmac. exp. Ther.* **89**, 64–76.

Rosell, S., Kopin, I. J. & Axelrod, J. (1963). Fate of H^3-norepinephrine in skeletal muscle before and following sympathetic stimulation. *Am. J. Physiol.* **205**, 317–321.

Ross, S. B. (1963). *In vivo* inactivation of catechol amines in mice. *Acta pharmac. tox.* **20**, 267–273.

Samorajski, T. & Marks, B. H. (1962). Localization of tritiated norepinephrine in mouse brain. *J. Histochem. Cytochem.* **10**, 392–399.

Schayer, R. W. (1951). Studies of the metabolism of β-C^{14}-DL-adrenaline. *J. biol. Chem.* **189**, 301–306.

Shore, P. A., Busfield, D. & Alpers, H. S. (1964). Binding and release of metaraminol: mechanism of norepinephrine depletion by α-methyl-*m*-tyrosine and related agents. *J. Pharmac. exp. Ther.* **146**, 194–199.

Snyder, S. H., Michaelson, I. A. & Musacchio, J. (1964). Purification of norepinephrine storage granules from rat heart. *Life Sciences* **3**, 965–970.

Stinson, R. H. (1961). Electrical stimulation of the sympathetic nerves of the isolated rabbit ear and the fate of the neurohormone released. *Can. J. Biochem. Physiol.* **39**, 309–316.

Stjärne, L. (1964). Studies of catecholamine uptake, storage and release mechanisms. *Acta physiol. scand.* **62**, Suppl. 228.

Stjärne, L. & Euler, U. S. von (1965). Stereospecificity of amine uptake mechanism in nerve granules. *J. Pharmac. exp. Ther.* **150**, 335–340.

Strömblad, B. C. R. (1959). Uptake of injected ^{14}C-adrenaline in denervated and in normally innervated submaxillary glands of the cat. *Br. J. Pharmac. Chemother.* **14**, 273–276.

Strömblad, B. C. R. (1961). Effects of insulin on adrenaline and noradrenaline content of some rat tissues. *Proc. Soc. exp. Biol. Med.* **108**, 345–347.

Strömblad, B. C. R. & Nickerson, M. (1961). Accumulation of epinephrine and norepinephrine by some rat tissues. *J. Pharmac. exp. Ther.* **134**, 154–159.

Thoenen, H., Hürlimann, A. & Haefely, W. (1964). The effect of sympathetic nerve stimulation on volume, vascular resistance and norepinephrine output in the isolated perfused spleen of the cat and its modification by cocaine. *J. Pharmac. exp. Ther.* **143**, 57–63.

Trendelenburg, U. & Crout, J. R. (1964). The norepinephrine stores of isolated atria of guinea-pigs treated with reserpine. *J. Pharmac. exp. Ther.* **145**, 151–161.

Udenfriend, S. & Wyngaarden, J. B. (1956). Precursors of adrenal epinephrine and norepinephrine *in vivo*. *Biochim. biophys. Acta* **20**, 48–52.

Westfall, T. C. (1965). Uptake and exchange of catecholamines in rat tissues after administration of D- and L-adrenaline. *Acta physiol. scand.* **63**, 336–342.

Whitby, L. G., Axelrod, J. & Weil-Malherbe, H. (1961). The fate of H^3-norepinêphrine in animals. *J. Pharmac. exp. Ther.* **132**, 193–201.

Wilbrandt, W. & Rosenberg, T. (1961). The concept of carrier transport and its corollaries in pharmacology. *Pharmac. Rev.* **13**, 109–183.

Wolfe, D. E. & Potter, L. T. (1963). Localization of norepinephrine in the atrial myocardium. *Anat. Rec.* **145**, 301.

Wolfe, D. E., Potter, L. T., Richardson, K. C. & Axelrod, J. (1962). Localizing tritiated norepinephrine in sympathetic axons by electron micrograph autoradiography. *Science* **138**, 440.

Wurtman, R. J., Kopin, I. J., Horst, D. W. & Fischer, J. E. (1964). Epinephrine and organ blood flow: effects of hyperthyroidism, cocaine and sympathetic denervation. *Am. J. Physiol.* **207**, 1247–1250.

Wylie, D. W., Archer, S. & Arnold, A. (1960). Augmentation of pharmacological properties of catecholamines by *O*-methyl transferase inhibitors. *J. Pharmac. exp. Ther.* **130**, 239–244.

Zaimis, E., Berk, L. & Callingham, B. (1965). Morphological, biochemical and functional changes in the sympathetic nervous system of rats treated with NGF-antiserum. *Nature, Lond.* **206**, 1221–1222.

8

SOME ASPECTS OF THE PHARMACOLOGY OF DRUGS WHICH INTERFERE WITH NORADRENALINE UPTAKE AND STORAGE

The discovery that many drugs can influence the uptake and storage of noradrenaline in sympathetic nerves has aroused great interest in recent years. Indeed 'there is now an almost compulsive desire to demonstrate, for all drugs which affect structures innervated by sympathetic nerve fibres, a drug effect in terms of either uptake or storage of adrenergic transmitter' (Zaimis, 1964). The interest in this field of pharmacology has been understandably heightened by the possibility that many of the drugs which have such actions on peripheral adrenergic mechanisms may prove to act similarly on adrenergic mechanisms in the central nervous system.

Inevitably in research which has involved scientists from many disciplines, misunderstandings exist in this field. An important source of confusion has been a widespread failure to define and to distinguish between the terms 'uptake' and 'storage'. 'Uptake' refers only to the process whereby exogenous catecholamines gain entry into sympathetic nerve fibres. 'Storage' refers to the processes by which endogenous and exogenous noradrenaline are retained in a stable form within sympathetic nerve fibres. As discussed in the previous chapter, the uptake of catecholamines may involve a specific transport mechanism in the axonal membrane of sympathetic nerves, whilst the storage of catecholamines in sympathetic nerves involves a binding of the catecholamines in intraneuronal storage particles. In many studies, however, 'uptake' has been equated with 'storage'; the 'uptake' of noradrenaline has often been measured as the amount of accumulated material remaining in a tissue many minutes or hours after the original uptake has occurred. Drug effects on uptake and storage can thus be confused. Drugs which affect the uptake of catecholamines can only

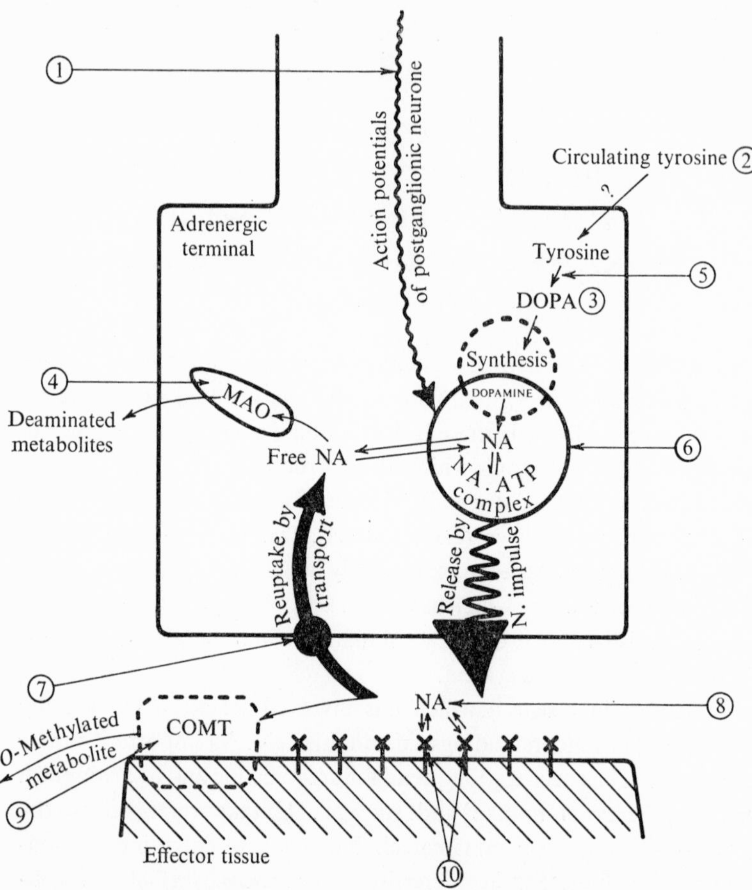

Fig. 8.1. Possible sites of action of drugs interfering with adrenergic mechanisms.

(1) Action potentials propagate to the terminals of sympathetic postganglionic neurones. Propagation in terminal regions may be blocked by adrenergic neurone-blocking drugs such as bretylium and guanethidine. Propagation may also be blocked at the synapse between pre- and postganglionic neurones by ganglion blocking drugs such as pempidine and chlorisondamine.

(2) Circulating tyrosine is the probable precursor used by adrenergic terminals for the biosynthesis of noradrenaline, circulating tyrosine enters the adrenergic terminal possibly by a carrier-mediated transport process (?). Structurally related drugs such as α-methyl-*m*-tyrosine can take the place of tyrosine and be converted by the biosynthetic process into false adrenergic neurotransmitters.

(3) False neurotransmitters can also be synthesized from drugs which are related to other intermediates in noradrenaline biosynthesis, for instance α-methyl DOPA can take the place of DOPA. *Continued on facing page.*

do so by altering the rate of the initial entry of extracellular amines into sympathetic nerves. The total amount of noradrenaline finally accumulated and stored might be unaffected by a partial inhibition of uptake and the drug effect could therefore go unnoticed. Conversely, a drug may prevent the normal intraneuronal storage of catecholamines without influencing the rate of uptake. In this case little or none of the accumulated catecholamine would remain in the tissue stores, and by conventional methods used for measuring 'uptake' it would be concluded that the drug inhibited catecholamine uptake. The outstanding example of a drug to which this applies is reserpine, which will be discussed in more detail below. Confusion of a similar type exists regarding the effects of monoamine oxidase inhibitors on noradrenaline uptake. Axelrod, Hertting & Patrick (1961) showed that in rats treated with inhibitors of monoamine oxidase the initial uptake of H^3-noradrenaline into the heart was unaffected. However, in the treated animals the accumulated noradrenaline disappeared from the tissue more slowly than in normal animals. Thus, if the 'uptake' of noradrenaline was estimated merely by measuring the amount of exogenous noradrenaline retained in the tissue at some time after the original uptake

---

(4) Free noradrenaline in the axoplasm is destroyed by monoamine oxidase, situated in intraneuronal mitochondria. This enzyme can be inhibited by a wide range of MAO inhibitors such as pheniprazine, nialamide or iproniazid.

(5) The rate-limiting step in noradrenaline biosynthesis, tyrosine hydroxylase, can be inhibited by drugs such as α-methyl-*p*-tyrosine or 3-iodotyrosine.

(6) The storage of noradrenaline in intraneuronal storage particles is prevented by drugs such as reserpine.

(7) The re-uptake of noradrenaline released by nerve impulses is effected by a membrane transport process which can be inhibited by many drugs, including cocaine, desipramine and many sympathomimetic amines.

(8) Many drugs can also be taken up into the adrenergic neurone by acting as substrates for this transport process. Once inside the neurone, drugs may displace noradrenaline from intraneuronal stores (indirect acting sympathomimetics), and may further take the place of noradrenaline by acting as false neurotransmitters (metaraminol, octopamine, α-methyl noradrenaline).

(9) The extraneuronal metabolism of noradrenaline by catechol-*O*-methyl transferase can be inhibited by pyrogallol and tropolones.

(10) The interaction of noradrenaline with α- and β-adrenergic receptors can be blocked by receptor blocking drugs such as phenoxybenzamine, phentolamine, DCI and pronethalol. The actions of noradrenaline on adrenergic receptors can be mimicked by direct-acting sympathomimetic amines such as adrenaline or synephrine.

of noradrenaline, it would appear that monoamine oxidase inhibition actually increased the 'uptake'.

In the case of drugs which interfere with the storage of noradrenaline there is a similar tendency to oversimplify very complex and diverse pharmacological effects. The great majority of chemical studies have merely established that certain drugs can influence the storage levels of noradrenaline in sympathetic nerves. The level of noradrenaline in sympathetic nerves, although relatively constant under normal conditions, cannot be regarded as a reflection of a static binding of noradrenaline in a physico-chemical sense. Rather the storage level of transmitter reflects a balance between the dynamic processes of transmitter synthesis and utilization. It is clear that drugs which alter the storage levels of noradrenaline may have very diverse sites of action. Thus drugs could affect the level of stored transmitter by interacting with any of the enzymes involved in the biosynthesis of the transmitter, or with the enzymes involved in the metabolic degradation of noradrenaline; alternatively drugs may alter the properties of the storage sites in the sympathetic nerves or may prevent the normal release of transmitter in response to sympathetic nerve stimulation.

A general criticism has been the failure to view the adrenergic synapse as a highly complex integrated functional unit. Within this unit the following systems must be considered: the presynaptic nerve terminal, with its complex anatomical structure including specialized intracellular storage particles and the biochemical machinery concerned with the synthesis and degradation of the neurotransmitter, and the biochemical or biophysical machinery for the release and reuptake of the transmitter. Also included in the synapse is the specialized region of the postsynaptic cell which is in intimate contact with the presynaptic nerve terminal; this part of the adrenergic synapse has received little attention in recent years with our current obsession in interpreting drug effects in terms of actions on the uptake and storage of noradrenaline in the presynaptic terminal (for an excellent review of the pharmacology of adrenergic receptors see Nickerson, 1965). Because of the many functional units which make up the whole structure, great caution is needed in extrapolating drug effects observed on a single functional unit to explain drug actions on the system as a whole. In evaluating drug actions on adrenergic transmission it should be remembered that noradrenaline is a key molecule which interacts with many chemical sites (Fig. 8.1). It should not be surprising that drug molecules which resemble noradrenaline in one or more aspects of their structure should also interact with more than one of these sites. Thus drugs with multiple sites of action on adrenergic mechanisms are more the rule than the exception.

In the present chapter only a small section of the now very considerable literature on the actions of drugs on noradrenaline uptake and storage can be considered. Because of the diverse actions of many of these drugs it is impossible to make a sharp distinction between drugs which interfere with noradrenaline uptake and drugs which interfere with noradrenaline storage, since many drugs may have both types of action.

A. DRUGS WHICH INHIBIT THE UPTAKE OF NORADRENALINE

1. Cocaine

The inhibition of catecholamine uptake by cocaine may be taken as an example of a drug which seems to act fairly selectively on adrenergic transmission. There is little evidence that cocaine has any marked effects on noradrenaline storage or metabolism in sympathetic nerves, while there is good evidence that this drug potently inhibits the reuptake of the transmitter.

There have been many theories to explain the actions of cocaine in potentiating the actions of noradrenaline or sympathetic nerve stimulation (for review see Furchgott, 1955). The most generally accepted current hypothesis is that cocaine acts by inhibiting the normally rapid inactivation of catecholamines by nerve uptake. When the uptake system is blocked higher concentrations of noradrenaline or other catecholamines are available in the vicinity of adrenergic receptors to produce exaggerated and prolonged pharmacological responses. This hypothesis has been criticized on the grounds that it cannot explain why cocaine evokes a greater response to noradrenaline in, for instance, an isolated tissue exposed to noradrenaline in an organ bath. The argument being that the tissue is exposed to a fixed concentration in the bath fluid and that this concentration cannot be appreciably altered by the small amounts of noradrenaline removed by uptake into the tissue. However, it is not the concentration of noradrenaline in the bath which determines the pharmacological response; it is the concentration of noradrenaline in the minute volume of extracellular fluid in immediate contact with adrenergic receptors which is important. It seems reasonable to suppose that this local concentration of noradrenaline could be markedly influenced by the operation of the catecholamine uptake process in adrenergic nerve terminals immediately adjacent to the postsynaptic effector sites. The 'nerve uptake hypothesis' was proposed by Macmillan (1959), Whitby, Hertting & Axelrod (1960) and Muscholl (1961). Direct evidence for an inhibition of noradrenaline uptake by cocaine was provided by Whitby *et al.* (1960) and by Muscholl (1961). Whitby *et al.* (1960) also showed that cocaine considerably delayed the disappearance of injected noradrenaline from the circulation (as previously reported by Trendelenburg, 1959) and the drug simultaneously inhibited the uptake of circulating noradrenaline into various peripheral tissues. The ability of cocaine to inhibit noradrenaline uptake was confirmed in tissue slices incubated with H^3-noradrenaline by Dengler, Spiegel & Titus (1961), in the isolated perfused rabbit and rat hearts by Lindmar & Muscholl (1962, 1964), and in the perfused cat spleen by Thoenen, Hürlimann & Haefely (1964*a*) and Gillespie & Kirpekar (1965). In the isolated perfused rat heart cocaine was found to be a potent inhibitor of the uptake of H^3-noradrenaline, producing a 50% inhibition of noradrenaline uptake at a drug concentration of $3{\cdot}8 \times 10^{-7}$ M (Iversen, 1965*a*). Muscholl (1961) showed that other related drugs such as tetracaine, which do not potentiate the actions of catecholamines, also failed to block the uptake of noradrenaline.

These reports have strongly supported the view that cocaine potentiates the actions of catecholamines by virtue of its ability to inhibit noradrenaline uptake. Indeed, such a potentiation of the actions of catecholamines is a property which all inhibitors of noradrenaline uptake may be expected to share. In addition to potentiating the actions of exogenously administered catecholamines, inhibitors of uptake should also potentiate the actions of noradrenaline released in response to sympathetic nerve stimulation. Such a potentiation has been demonstrated for cocaine (Trendelenburg, 1959; Haefely, Hürlimann & Thoenen, 1964; Huković, 1959). In addition to the potentiation of the physiological effects of noradrenaline released by nerve stimulation, it might be expected that cocaine would promote the overflow of transmitter on stimulation of sympathetic nerves. There have been several attempts to demonstrate that after cocaine treatment more noradrenaline appears in the venous effluent from a tissue on stimulation of the sympathetic nerve supply. In the cat spleen preparation Trendelenburg (1959) and Blakeley, Brown & Ferry (1963) could not detect such an increased outflow of noradrenaline, and Kirpekar & Cervoni (1963) reported only a small increase in the overflow of noradrenaline during nerve stimulation in cocaine-treated animals. However, other experiments have indicated that in the cat spleen stimulated at very low frequencies cocaine does produce a considerable increase in the overflow of noradrenaline (Table 8.2, p. 157) (Thoenen *et al.* 1964*a*). Cocaine produced a similar increase in the overflow of noradrenaline in the cat nictitating membrane and in the rabbit eye on sympathetic nerve stimulation (Haefely *et al.* 1964; Eakins & Eakins, 1964). In the isolated rabbit heart cocaine more than doubled the overflow of noradrenaline on stimulation of the sympathetic nerve supply (Huković & Muscholl, 1962).

Cocaine is apparently relatively specific in its actions on catecholamine uptake. It is not an inhibitor of the biosynthesis of noradrenaline or of catechol-*O*-methyl transferase (Holtz, Osswald & Stock, 1960) and it is only a very weak inhibitor of monoamine oxidase (Brown & Hey, 1956). Cocaine has little effect on the tissue storage of noradrenaline, since the administration of cocaine fails to produce any change in the noradrenaline content of tissues (Muscholl, 1961). There is some evidence that the inhibition of noradrenaline uptake by cocaine is readily reversible; Muscholl (1961) found that the inhibition of noradrenaline uptake produced by cocaine in the rat heart *in vivo* depended on the relative size of the doses of noradrenaline and cocaine used. In the isolated rat heart the dose response curve for the inhibition of noradrenaline uptake by cocaine was shifted to the right when the concentration of noradrenaline in the perfusing medium was increased (Iversen, 1963). Furchgott, Kirpekar, Rieker & Schwab (1963), Farmer & Petch (1963) and Farrant (1963) have also concluded from pharmacological evidence that cocaine and noradrenaline compete for a common uptake site in sympathetic nerves.

Histochemical studies of noradrenaline uptake in the rat iris have provided direct evidence for the view that cocaine blocks noradrenaline uptake by interfering with the specific uptake mechanism in the axonal membrane of sympathetic nerve fibres (Hillarp & Malmfors, 1964; Malmfors, 1965). In these studies it was shown that cocaine inhibited the accumulation of noradrenaline in both terminal and preterminal fibres of the adrenergic plexus in the iris. Since storage particles are either lacking or very sparsely distributed in preterminal regions of the sympathetic fibres the authors concluded that cocaine could not be inhibiting uptake by interfering with the accumulation of noradrenaline in intraneuronal storage particles, but must act at the level of the axonal membrane. Hillarp & Malmfors (1964) also showed that cocaine could block the uptake of α-methyl noradrenaline in the sympathetic fibres of the rat iris. Muscholl & Weber (1965) have also shown that cocaine can inhibit the uptake of α-methyl noradrenaline in the isolated perfused rat heart. Cocaine in addition inhibits the uptake of adrenaline (Hardman & Mayer, 1965), metaraminol (Carlsson & Waldeck, 1965) and α-methyl tyramine (Iversen, 1966). These reports support the view that these and other amines structurally related to noradrenaline may be accumulated by a common uptake mechanism in sympathetic nerves. In contrast to these reports it has been found that cocaine fails to inhibit the uptake of catecholamines in the rat uterus (Wurtman, Kopin & Axelrod, 1963) or in sympathetic ganglia (Fischer & Snyder, 1966), suggesting that in these tissues there is either a qualitatively different uptake mechanism or that the drug fails to penetrate to the uptake sites.

2. Adrenergic blocking agents

The uptake of H^3-noradrenaline into peripheral tissues was reported to be inhibited by phenoxybenzamine and dichloroisoprenaline (DCI) by Axelrod, Hertting & Potter (1962). Dengler *et al.* (1961) failed to confirm the inhibition of noradrenaline uptake by phenoxybenzamine but found ergotamine to be effective in inhibiting noradrenaline uptake. Farrant, Harvey & Pennefather (1964) and Lindmar & Muscholl (1964) confirmed the ability of phenoxybenzamine and DCI to inhibit noradrenaline uptake, the latter authors also reported that the β-receptor blocking agent pronethalol acted as an inhibitor of noradrenaline uptake. In the isolated rat heart

Table 8.1. *The inhibition of noradrenaline Uptake$_1$ by drugs in the isolated rat heart*

(From Iversen, 1965*a*.)

Drug	Drug concentration (M×)	Inhibition of noradrenaline uptake (%)	Drug ID 50 (M×)
Desipramine	1×10^{-8}	43·5	
Desipramine	1×10^{-7}	81·0	$1{\cdot}3\times10^{-8}$
Desipramine	1×10^{-6}	92·5	
Imipramine	1×10^{-7}	45·5	$9{\cdot}0\times10^{-8}$
Imipramine	1×10^{-6}	75·0	
bis-Desipramine*	1×10^{-7}	37·0	—
Reserpine	1×10^{-6}	73·0	—
Chlorpromazine	1×10^{-5}	88·5	—
Cocaine	1×10^{-7}	25·0	
Cocaine	1×10^{-6}	68·0	$3{\cdot}8\times10^{-7}$
Cocaine	1×10^{-5}	95·0	
Guanethidine	2×10^{-6}	41·0	$3{\cdot}3\times10^{-6}$
Guanethidine	2×10^{-5}	78·5	
Bretylium	2×10^{-6}	10·0	$1{\cdot}4\times10^{-5}$
Bretylium	2×10^{-5}	60·0	
Adrenergic blocking agents			
Phenoxybenzamine	1×10^{-5}	91·5	—
3,4-Dichloroisoprenaline	2×10^{-6}	50·5	—
Pronethalol	5×10^{-6}	36·0	—
Phentolamine	1×10^{-5}	66·0	—
Dibenamine†	1×10^{-5}	34·0	—
Monoamine oxidase inhibitors			
Tranylcypromine	1×10^{-5}	88·0	—
Harmine‡	1×10^{-5}	79·5	—
Phenelzine	1×10^{-5}	69·0	—
Isocarboxazide	1×10^{-5}	14·5	—
Nialamide	1×10^{-5}	3·5	—
Pargyline	1×10^{-5}	Nil	—
Iproniazi	1×10^{-5}	Nil	—

* 5-(3-Aminopropyl)-10,11-dihydro-5H-dibenz(b,f)azepine.
† *N,N*-Dibenzyl-β-chloroethylamine.
‡ 7-Methoxy-1-methyl-9-pyrid(3,4-b)indole.

phenoxybenzamine, DCI, pronethalol, phentolamine and dibenamine were found to act as inhibitors of H^3-noradrenaline uptake (Table 8.1; and Iversen, 1965*a*). Gillespie & Kirpekar (1965) have also confirmed the ability of phentolamine, hydergine and phenoxybenzamine to inhibit the removal of infused noradrenaline by the cat spleen.

The ability of adrenergic blocking agents to inhibit noradrenaline uptake is particularly important because of the confusion which has arisen over the dual actions of these drugs. There seems little doubt that the uptake of noradrenaline occurs into sympathetic nerves and not into adrenergic receptor sites. Furthermore, there are many drugs, such as cocaine, which inhibit the uptake of noradrenaline without having any adrenergic blocking activity. Before it was known that the uptake of noradrenaline occurred into sympathetic nerves, it was quite reasonably supposed that the receptor blocking agents interfered with the uptake or binding of noradrenaline at adrenergic receptors (Brown & Gillespie, 1957; Blakeley *et al.* 1963; Kirpekar & Cervoni, 1963).

The adrenergic blocking agents have most of the properties which are predicted as a consequence of their ability to inhibit noradrenaline uptake. Thus, phenoxybenzamine potentiates the actions of noradrenaline and adrenaline on rabbit atria (Stafford, 1963). Phenoxybenzamine and other blocking agents greatly increase the outflow of noradrenaline from the cat spleen stimulated at low frequencies (Brown & Gillespie, 1957; Brown, 1965; Thoenen *et al.* 1964*b*) (Fig. 6.2). Phenoxybenzamine also promotes the overflow of H^3-noradrenaline from skeletal muscle on stimulation of the sympathetic nerve supply (Rosell, Kopin & Axelrod, 1963), and potentiates the effects of noradrenaline released by sympathetic nerve stimulation in rabbit atria (Huković, 1959). DCI promoted the outflow of noradrenaline from the rabbit heart on stimulation of the sympathetic nerve supply (Huković & Muscholl, 1962). The increased outflow of transmitter from tissues treated with adrenergic blocking agents (Table 8.2) may be the result not only of the effect of these drugs in inhibiting transmitter re-uptake but also of their action in blocking the local vasoconstriction normally caused by noradrenaline release. As discussed in the previous chapter the efficiency of the re-uptake mechanism for the inactivation of noradrenaline is highly dependent on the rate of blood flow through the tissue, and when adrenergic blocking agents are administered blood flow may be abnormally high (Hertting & Schiefthaler, 1963; Rosell *et al.* 1963). In the case of adrenergic blocking agents such as phentolamine, which are only weakly effective as inhibitors of uptake, almost the whole of their effects on transmitter overflow may be related to their effects on blood flow (Hertting, 1965).

Table 8.2. *Effects of various drugs on the overflow of noradrenaline from the perfused cat spleen during stimulation of the sympathetic nerve supply*

(From Thoenen, Hürlimann & Haefely, 1964*a, b, c.*)

Drug	Noradrenaline output during 10 sec stimulation period at frequency of 6/sec (ng ± S.E.M.)	
	Control	Drug-treated
Cocaine	40·6 ± 4·0	70·6 ± 6·0
Phenoxybenzamine	47·5 ± 6·0	115·0 ± 15·7
Phentolamine	52·7 ± 4·7	87·9 ± 8·2
Azapetine	62·8 ± 5·3	84·9 ± 12·4
Imipramine	38·3 ± 4·5	74·3 ± 8·3

The perfused cat spleen was stimulated at intervals of 8 min for 10 sec periods of supramaximal stimulation of the splenic nerve at a frequency of 6/sec. The output of noradrenaline into the perfusing medium was assayed biologically. Drugs were added to the perfusion medium and administered at a rate of 3 μg/min. Values are the means for 7–8 experiments ± S.E.M.

There have been reports that the prolonged administration of phenoxybenzamine causes a lowering in the noradrenaline content of peripheral tissues (Schapiro, 1958; Farrant *et al.* 1964). It is possible that such an effect could be due to a prolonged inhibition of noradrenaline uptake; under such conditions noradrenaline may be released from the tissues by sympathetic nerve activity more rapidly than it can be resynthesized, the normal return of the transmitter to the tissue stores being prevented. Blakeley, Brown & Geffen (1964) demonstrated that in the cat spleen in the presence of phenoxybenzamine successive trains of stimuli released progressively smaller amounts of noradrenaline; presumably because the noradrenaline available for release was rapidly exhausted in the absence of transmitter re-uptake.

3. Sympathomimetic amines

The sympathomimetic amines tyramine, amphetamine and ephedrine were reported to act as inhibitors of noradrenaline uptake (Axelrod & Tomchick, 1960; Axelrod, Whitby & Hertting, 1961; Hertting, Axelrod & Whitby, 1961; Dengler *et al.* 1961). These findings prompted a detailed study of the inhibition of noradrenaline uptake by amines structurally related to noradrenaline. In these experiments the various amines were tested as inhibitors of the uptake of low concentrations of labelled noradrenaline in the isolated perfused rat heart and for each amine the concentration producing a 50% inhibition of noradrenaline uptake (ID 50) was determined. The reciprocal of this concentration was taken to represent an estimate of the affinity of the various amines for the uptake site in the sympathetic nerves (Iversen, 1964; Burgen & Iversen, 1965). Tyramine and other structurally related amines produced an immediate inhibition of noradrenaline uptake, probably by competing with noradrenaline for uptake sites. The results of these experiments are summarized in Tables 8.3–8.9. The tables are arranged to illustrate the effects of various structural changes on the affinity of the drugs for the noradrenaline uptake process. Almost all of the amines tested had some activity as inhibitors of noradrenaline uptake. From the results obtained it was possible to describe several very consistent structure–activity relationships which define the structural specificity of the catecholamine uptake site in the sympathetic nerves of the rat heart (Burgen & Iversen, 1965). It should be emphasized that the method used does not distinguish between drugs which inhibit noradrenaline uptake but are not transported into the tissue by the uptake process, and drugs which inhibit noradrenaline uptake by acting as competitive substrates for uptake. As described in the previous chapter, several of the sympathomimetic amines which were potent inhibitors of noradrenaline uptake have subsequently been found to act as substrates for the uptake process, but this may not necessarily prove to be a general rule. The structure–activity relationships among the sympathomimetic amines for inhibition of noradrenaline uptake can be summarized as follows:

(*a*) β-Hydroxylation produced a decreased affinity for the uptake site. In such compounds the L-enantiomer had a higher affinity than the D-enantiomer.

(*b*) α-Methylation resulted in a considerable increase in affinity for the uptake site. This effect was also stereochemically specific, in this case the D-enantiomer had considerably more activity than the L-enantiomer.

(*c*) Phenolic hydroxyl groups enhanced the affinity for the uptake site, *para*- and *meta*- substitutions having approximately equal effects. The optimal structure was the 3,4-dihydroxyphenyl group.

(*d*) *N*-Substitution decreased the affinity of the drug for the uptake site. This effect was dependent on the size of the *N*-substituent, bulky substituents having correspondingly greater effects in depressing affinity.

(*e*) *O*-Methylation of phenolic hydroxyl groups produced a striking decrease in affinity for the uptake site. *meta*-Methoxy compounds had considerably lower affinities than the corresponding *para*-methoxy compounds.

(*f*) The phenylethylamine structure could be replaced with saturated five- or six-membered ring structures as in propylhexedrine and cyclopentamine without producing a marked decrease in the affinity for uptake. Even the long chain aliphatic amine tuamine and the indoleamine serotonin had appreciable affinities for the uptake site.

Muscholl & Weber (1965) measured the potency of various sympathomimetic amines as inhibitors of the uptake of α-methyl noradrenaline in the isolated perfused rat heart. The results of this investigation are in good agreement with those described above; introduction of phenolic hydroxyl groups in the meta-position increased inhibitory potency, β-hydroxylation decreased potency, α-methylation increased potency and the effects were stereochemically specific. Ross & Renyi (1965) tested various sympathomimetic amines as inhibitors of the accumulation of tritiated noradrenaline in mouse brain slices incubated *in vitro*. The results of this study are somewhat different from those reported in the rat heart; the effects of adrenaline and noradrenaline as inhibitors of H^3-noradrenaline accumulation were not stereochemically specific and no clear structure–activity relationships were apparent.

The finding that many sympathomimetic amines can act as potent inhibitors of noradrenaline uptake has interesting pharmacological implications. In the first place, as with other inhibitors of uptake, the sympathomimetic amines may be expected to potentiate the actions of noradrenaline and other catecholamines. The actions of ephedrine and tyramine in potentiating the effects of exogenous catecholamines have been described (Gaddum & Kwiatkowski, 1938; Furchgott *et al.*

Table 8.3. *Inhibition of noradrenaline uptake by sympathomimetic amines in the rat heart (Uptake$_1$)*

(From Burgen & Iversen, 1965.)

Drug	ID 50 (drug concentration producing 50 % inhibition of noradrenaline uptake)	Relative affinity for uptake site (phenethylamine = 100)
(−)-Metaraminol	$7{\cdot}6 \times 10^{-8}$	1440
Dopamine	$1{\cdot}7 \times 10^{-7}$	650
(±)-α-Methyldopamine	$1{\cdot}8 \times 10^{-7}$	610
(+)-Amphetamine	$1{\cdot}8 \times 10^{-7}$	610
(±)-Hydroxyamphetamine (paredrine)	$1{\cdot}8 \times 10^{-7}$	610
(−)-α-Methylnoradrenaline (cobefrin)	$2{\cdot}0 \times 10^{-7}$	550
(−)-Noradrenaline	$2{\cdot}7 \times 10^{-7}$	407
(±)-α-Methylnoradrenaline	$4{\cdot}2 \times 10^{-7}$	260
Tyramine	$4{\cdot}5 \times 10^{-7}$	245
(±)-Amphetamine	$4{\cdot}6 \times 10^{-7}$	240
m-Tyramine	$5{\cdot}1 \times 10^{-7}$	215
(+)-Methamphetamine	$6{\cdot}7 \times 10^{-7}$	165
(±)-Noradrenaline	$6{\cdot}7 \times 10^{-7}$	165
(±)-Prenylamine (segontin)	$7{\cdot}4 \times 10^{-7}$	149
N-Methyldopamine (epinine)	$7{\cdot}6 \times 10^{-7}$	145
(±)-Nylidrin (arlidin)	$8{\cdot}5 \times 10^{-7}$	130
(±)-Propylhexedrine	$8{\cdot}5 \times 10^{-7}$	130
(−)-Adrenaline	$1{\cdot}0 \times 10^{-6}$	110
Mephentermine	$1{\cdot}0 \times 10^{-6}$	110
Phenethylamine	$1{\cdot}1 \times 10^{-6}$	100
(±)-Norsynephrine (octopamine)	$1{\cdot}3 \times 10^{-6}$	85
Tranylcypromine	$1{\cdot}3 \times 10^{-6}$	85
(+)-Noradrenaline	$1{\cdot}4 \times 10^{-6}$	79
(±)-Cyclopentamine (clopane)	$1{\cdot}4 \times 10^{-6}$	79
Noradrenalone (3,4-dihydroxy-α-methyl-aminoacetophenone)	$1{\cdot}5 \times 10^{-6}$	73
β-Naphthylethylamine	$1{\cdot}5 \times 10^{-6}$	73
(±)-Adrenaline	$1{\cdot}6 \times 10^{-6}$	70
(±)-Phenylpropanolamine (propadrine)	$2{\cdot}0 \times 10^{-6}$	55
(±)-3,4-Dichloroisoprenaline	$2{\cdot}0 \times 10^{-6}$	55
(−)-Ephedrine	$2{\cdot}2 \times 10^{-6}$	50
N-Dimethyl-*p*-tyramine (hordenine)	$2{\cdot}5 \times 10^{-6}$	45
(±)-*N*-Ethylnoradrenaline	$3{\cdot}2 \times 10^{-6}$	34
(−)-Amphetamine	$3{\cdot}7 \times 10^{-6}$	30
Phenelzine	$3{\cdot}8 \times 10^{-6}$	29
(±)-β-Hydroxyphenethylamine	$4{\cdot}8 \times 10^{-6}$	23
(−)-Phenylephrine (neosynephrine)	$5{\cdot}6 \times 10^{-6}$	20

Table 8.3. (*cont.*)

Drug	ID 50 (drug concentration producing 50% inhibition of noradrenaline uptake)	Relative affinity for uptake site (phenethylamine = 100)
1-Methylhexylamine (tuamine)	$5{\cdot}6 \times 10^{-6}$	20
(±)-*N*-Ethylnoradrenaline	$9{\cdot}2 \times 10^{-6}$	12
p-Methoxyphenethylamine	$1{\cdot}0 \times 10^{-5}$	11
(±)-Methoxyphrenamine (orthoxine)	$1{\cdot}1 \times 10^{-5}$	10
(±)-Synephrine	$1{\cdot}2 \times 10^{-5}$	9
5-Hydroxytryptamine	$2{\cdot}0 \times 10^{-5}$	5·5
(±)-Isoprenaline	$2{\cdot}5 \times 10^{-5}$	4·5
(±)-*N*-Butylnoradrenaline	$3{\cdot}5 \times 10^{-5}$	3·2
(±)-*N*-Isobutylnoradrenaline	$4{\cdot}0 \times 10^{-5}$	2·6
(±)-3-*O*-Methyladrenaline (metanephrine)	$4{\cdot}3 \times 10^{-5}$	2·6
(−)-DOPA	$6{\cdot}0 \times 10^{-5}$	1·2
(±)-3-*O*-Methylnoradrenaline (normetanephrine)	$2{\cdot}0 \times 10^{-4}$	0·55
2,4,5-Trihydroxyphenethylamine	$> 2{\cdot}0 \times 10^{-4}$	< 0·55
3,4-Dimethoxyphenethylamine	$2{\cdot}0 \times 10^{-4}$	0·55
(±)-Methoxamine (vasoxyl)	$1{\cdot}0 \times 10^{-3}$	0·11
3,4,5-Trimethoxyphenethylamine (mescaline)	$1{\cdot}5 \times 10^{-2}$	0·007

1963). The dual action of amines such as tyramine which release noradrenaline from tissue stores and simultaneously inhibit the re-uptake of the released catecholamine can explain why the very small amounts of noradrenaline released by such amines can have such potent effects on the effector tissue (Lindmar & Muscholl, 1961, 1965; Weiner, Draskóczy & Burack, 1962). A second possibility suggested by these results is that many of the amines which have high affinities for the uptake site in sympathetic nerves may themselves be taken up into tissues. It is known that adrenaline, metaraminol, α-methyl noradrenaline and α-methyl tyramine are indeed rapidly taken up into tissues (chapter 7). In the case of indirectly acting sympathomimetic amines such as tyramine, an uptake into sympathetic nerves may be a necessary prelude to the subsequent displacement of noradrenaline from storage sites inside the nerve. This would explain, for instance, why cocaine and other inhibitors of the catecholamine uptake process are

Table 8.4. *Inhibition of noradrenaline Uptake$_1$ by related amines. I. Effects of phenolic hydroxyl groups*

(From Burgen & Iversen, 1965.)

CH—CH—NH

Drug	(a)	(b)	(c)	(d)	(e)	ID 50	Relative affinity for uptake
Phenethylamine*	H	H	H	H	H	1.1×10^{-6}	100
p-Tyramine	**OH**	H	H	H	H	4.5×10^{-7}	245
m-Tyramine	H	**OH**	H	H	H	5.1×10^{-7}	215
Dopamine	**OH**	**OH**	H	H	H	1.7×10^{-7}	650
DL-Phenyl-propanolamine*	H	H	OH	CH_3	H	2.0×10^{-6}	55
L-Metaraminol	H	**OH**	OH	CH_3	H	7.6×10^{-8}	1440
L-Phenylephrine*	H	**OH**	OH	H	CH_3	5.6×10^{-6}	20
DL-Synephrine	**OH**	H	OH	H	CH_3	1.2×10^{-5}	9
DL-Adrenaline	**OH**	**OH**	OH	H	CH_3	1.4×10^{-6}	78
L-Adrenaline	**OH**	**OH**	OH	H	CH_3	1.0×10^{-6}	110
DL-Amphetamine*	H	H	H	CH_3	H	4.6×10^{-7}	240
DL-Hydroxyamphet-amine	**OH**	H	H	CH_3	H	1.8×10^{-7}	610
DL-α-Methyldopamine	**OH**	**OH**	H	CH_3	H	1.8×10^{-7}	610
DL-β-Hydroxy-phenethylamine	H	H	OH	H	H	4.8×10^{-6}	23
DL-Norsynephrine	**OH**	H	OH	H	H	1.3×10^{-6}	85
DL-Noradrenaline	**OH**	**OH**	OH	H	H	6.7×10^{-7}	164

* Unsubstituted reference molecule.

able to block the actions of tyramine and other indirectly acting amines. For direct-acting sympathomimetic amines, uptake into sympathetic nerves may serve as an inactivation mechanism, as it does for noradrenaline and adrenaline. The uptake of such amines would therefore be expected to decrease their pharmacological actions on the effector tissue. In support of this hypothesis it is well known that inhibitors of uptake such as cocaine potentiate the actions not only of noradrenaline but also of the directly acting sympathomimetic amines (Trendelenburg, Muskus, Fleming & Gomez, 1962*b*). Furthermore, cocaine and other inhibitors of uptake potentiate the actions of

Table 8.5. *Inhibition of noradrenaline Uptake*$_1$ *by related amines. II. Effects of α-methylation*

(From Burgen & Iversen, 1965.)

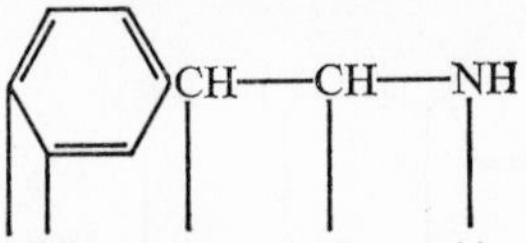

Drug	(a)	(b)	(c)	(d)	(e)	ID 50	Relative affinity for uptake
Phenethylamine*	H	H	H	H	H	$1{\cdot}1\times10^{-6}$	100
DL-Amphetamine	H	H	H	CH_3	H	$4{\cdot}6\times10^{-7}$	240
D-Amphetamine	H	H	H	CH_3	H	$1{\cdot}8\times10^{-7}$	610
L-Amphetamine	H	H	H	CH_3	H	$3{\cdot}7\times10^{-6}$	30
p-Tyramine*	OH	H	H	H	H	$4{\cdot}5\times10^{-7}$	245
DL-Hydroxy-amphetamine	OH	H	H	CH_3	H	$1{\cdot}8\times10^{-7}$	610
Dopamine*	OH	OH	H	H	H	$1{\cdot}7\times10^{-7}$	650
DL-α-Methyldopamine	OH	OH	H	CH_3	H	$1{\cdot}8\times10^{-7}$	610
L-Noradrenaline*	OH	OH	OH	H	H	$2{\cdot}7\times10^{-7}$	407
DL-Noradrenaline*	OH	OH	OH	H	H	$6{\cdot}7\times10^{-7}$	164
L-α-Methyl-noradrenaline	OH	OH	OH	CH_3	H	$2{\cdot}0\times10^{-7}$	550
DL-α-Methyl-noradrenaline	OH	OH	OH	CH_3	H	$4{\cdot}3\times10^{-7}$	256
D-Methamphetamine*	H	H	H	CH_3	CH_3	$6{\cdot}7\times10^{-7}$	165
Mephentermine	H	H	H	$(CH_3)_2$	CH_3	$1{\cdot}0\times10^{-6}$	110
DL-β-Hydroxy-phenethylamine*	H	H	OH	H	H	$4{\cdot}8\times10^{-6}$	23
DL-Phenyl-propanolamine	H	H	OH	CH_3	H	$2{\cdot}0\times10^{-6}$	55

* Unsubstituted reference molecule.

noradrenaline to a greater extent than they potentiate the actions of adrenaline, and they have practically no effect on the actions of isoprenaline (Stafford, 1963). These findings are consistent with results which indicate that noradrenaline had a higher affinity for the uptake mechanism than adrenaline and that isoprenaline had only a very low affinity. Since tissue uptake was thus most important for noradrenaline, less important for adrenaline and not important at all for isoprenaline, the effects of these amines are potentiated to varying degrees by the inhibition of uptake.

Table 8.6. *Inhibition of noradrenaline Uptake$_1$ by related amines. III. Effects of β-hydroxylation*

(From Burgen & Iversen, 1965.)

Drug	(a)	(b)	(c)	(d)	(e)	ID 50	Relative affinity for uptake
Phenethylamine*	H	H	H	H	H	$1 \cdot 1 \times 10^{-6}$	100
DL-β-Hydroxy-phenethylamine	H	H	**OH**	H	H	$4 \cdot 8 \times 10^{-6}$	23
Dopamine*	OH	OH	H	H	H	$1 \cdot 7 \times 10^{-7}$	650
L-Noradrenaline	OH	OH	**OH**	H	H	$2 \cdot 7 \times 10^{-7}$	407
DL-Noradrenaline	OH	OH	**OH**	H	H	$6 \cdot 7 \times 10^{-7}$	164
D-Noradrenaline	OH	OH	**OH**	H	H	$1 \cdot 4 \times 10^{-7}$	79
N-Methyldopamine*	OH	OH	H	H	CH_3	$7 \cdot 6 \times 10^{-7}$	145
DL-Adrenaline	OH	OH	**OH**	H	CH_3	$1 \cdot 4 \times 10^{-6}$	78
L-Adrenaline	OH	OH	**OH**	H	CH_3	$1 \cdot 0 \times 10^{-6}$	110
DL-Amphetamine*	H	H	H	CH_3	H	$4 \cdot 6 \times 10^{-7}$	240
DL-Phenyl-propanolamine	H	H	**OH**	CH_3	H	$2 \cdot 0 \times 10^{-6}$	55
D-Methamphetamine*	H	H	H	CH_3	CH_3	$6 \cdot 7 \times 10^{-7}$	165
L-Ephedrine	H	H	**OH**	CH_3	CH_3	$2 \cdot 2 \times 10^{-6}$	50
p-Tyramine*	OH	H	H	H	H	$4 \cdot 5 \times 10^{-7}$	245
DL-Norsynephrine	OH	H	**OH**	H	H	$1 \cdot 3 \times 10^{-6}$	85
DL-α-Methyl-dopamine*	OH	OH	H	CH_3	H	$1 \cdot 8 \times 10^{-7}$	610
DL-α-Methyl-noradrenaline	OH	OH	**OH**	CH_3	H	$4 \cdot 3 \times 10^{-7}$	256

* Unsubstituted reference molecule.

A further consequence of this aspect of the pharmacology of directly acting sympathomimetic amines is that it may be necessary to inhibit the uptake of these substances in order to measure their true potency at adrenergic receptors. The phenomenon of re-uptake, which is of varying quantitative importance for different amines, introduces an important variable in measurements of the direct activity of amines at adrenergic receptors. For instance, Trendelenburg (1966) has shown that the activities of the stereoisomers of adrenaline and noradrenaline

Table 8.7. *Inhibition of noradrenaline Uptake$_1$ by related amines.*
IV. Effects of N-*substitution*

(From Burgen & Iversen, 1965.)

Structure: ring–CH(c)–CH(d)–NH(e), ring substituents (a) (b)

Drug	(a)	(b)	(c)	(d)	(e)	ID 50	Relative affinity for uptake
L-Noradrenaline*	OH	OH	OH	H	H	$2\cdot7\times10^{-7}$	407
L-Adrenaline	OH	OH	OH	H	CH_3	$1\cdot0\times10^{-6}$	110
DL-Noradrenaline*	OH	OH	OH	H	H	$6\cdot7\times10^{-7}$	164
DL-Adrenaline	OH	OH	OH	H	CH_3	$1\cdot4\times10^{-6}$	78
DL-*N*-Ethylnoradrenaline	OH	OH	OH	H	C_2H_5	$4\cdot2\times10^{-6}$	12
DL-Isoprenaline	OH	OH	OH	H	$CH(CH_3)_2$	$2\cdot5\times10^{-5}$	4·5
DL-*N*-Propylnoradrenaline	OH	OH	OH	H	C_3H_7	$3\cdot7\times10^{-5}$	3·0
DL-*N*-Butylnoradrenaline	OH	OH	OH	H	C_4H_9	$3\cdot5\times10^{-5}$	3·2
DL-*N*-Isobutylnoradrenaline	OH	OH	OH	H	$CH_2CH(CH_3)_2$	$4\cdot0\times10^{-5}$	2·6
D-Amphetamine*	H	H	H	CH_3	H	$1\cdot8\times10^{-7}$	610
D-Methamphetamine	H	H	H	CH_3	CH_3	$6\cdot7\times10^{-7}$	165
p-Tyramine*	OH	H	H	H	H	$4\cdot5\times10^{-7}$	245
N-Dimethyl-*p*-tyramine	OH	H	H	H	$(CH_3)_2$	$2\cdot5\times10^{-6}$	44
Dopamine*	OH	OH	H	H	H	$1\cdot7\times10^{-7}$	650
Epinine	OH	OH	H	H	CH_3	$7\cdot6\times10^{-7}$	145
DL-Phenylpropanolamine*	H	H	OH	CH_3	H	$2\cdot0\times10^{-6}$	55
L-Ephedrine	H	H	OH	CH_3	CH_3	$2\cdot2\times10^{-6}$	50
DL-Norsynephrine*	OH	H	OH	H	H	$1\cdot3\times10^{-6}$	85
DL-Synephrine	OH	H	OH	H	CH_3	$1\cdot2\times10^{-5}$	9

* Unsubstituted reference compound.

Table 8.8. *Inhibition of noradrenaline Uptake$_1$ by related amines. V. Effects of* O-*methylation*

(From Burgen & Iversen, 1965.)

Drug	(*a*)	(*b*)	(*c*)	(*d*)	(*e*)	ID 50	Relative affinity for uptake
Phenethylamine*	H	H	H	H	H	$1{\cdot}1\times10^{-6}$	100
p-Methoxyphenethylamine	**OCH_3**	H	H	H	H	$1{\cdot}0\times10^{-5}$	11
3,4-Dimethoxyphenethylamine	**OCH_3**	**OCH_3**	H	H	H	$2{\cdot}0\times10^{-4}$	0·55
3,4,5-Trimethoxyphenethylamine	3,4,5-**OCH_3**		H	H	H	$1{\cdot}5\times10^{-2}$	0·007
DL-Noradrenaline*	OH	OH	OH	H	H	$6{\cdot}7\times10^{-7}$	164
3-*O*-Methylnoradrenaline	OH	**OCH_3**	OH	H	H	$2{\cdot}0\times10^{-4}$	0·55
DL-β-Hydroxyphenethylamine*	H	H	OH	CH_3	H	$2{\cdot}0\times10^{-6}$	55
DL-Methoxamine	2,5-**OCH_3**		OH	CH_3	H	$1{\cdot}0\times10^{-3}$	0·11
D-Methamphetamine*	H	H	H	CH_3	CH_3	$6{\cdot}7\times10^{-7}$	165
DL-Methoxyphenamine	2-**OCH_3**	H	H	CH_3	CH_3	$1{\cdot}1\times10^{-5}$	10

* Unsubstituted reference molecule.

Table 8.9. *Inhibition of noradrenaline Uptake₁ by related amines.*
VI. Effects of other substitutions

(From Burgen & Iversen, 1965.)

General structure: ring substituents (a), (b); side chain CH(c)—CH(d)—NH(e)

Drug	(a)	(b)	(c)	(d)	(e)	ID 50	Relative affinity for uptake
D-Methamphetamine*	H	H	H	CH_3	CH_3	$6{\cdot}7\times10^{-7}$	165
DL-Propylhexedrine	Cyclohexane H			CH_3	CH_3	$8{\cdot}5\times10^{-7}$	130
DL-Cyclopentamine	Cyclopentane H			CH_3	CH_3	$1{\cdot}4\times10^{-6}$	78
DL-Amphetamine*	H	H	H	CH_3	H	$4{\cdot}6\times10^{-7}$	240
DL-Nylidrin	OH	H	OH	CH_3	$CH(CH_3).CH_2CH_2C_6H_5$	$7{\cdot}9\times10^{-7}$	140
DL-Prenylamine	H	H	H	CH_3	$CH_2CH_2CH(C_6H_5)$	$7{\cdot}4\times10^{-7}$	149
1-Methylhexylamine	$CH_3CH_2CH_2CH_2CH_2CH(CH_3)NH_2$					$5{\cdot}6\times10^{-6}$	20
Dopamine*	OH	OH	H	H	H	$1{\cdot}7\times10^{-7}$	650
L-DOPA	OH	OH	H	COOH	H	$9{\cdot}4\times10^{-5}$	1·2
2,4,5-Trihydroxyphenethylamine	2,4,5-OH		H	H	H	Above 10^{-4}	Less than 1
DL-Isoprenaline*	OH	OH	OH	H	$CH(CH_3)_2$	$2{\cdot}5\times10^{-5}$	4·5
DL-3,4-Dichloroisoprenaline	Cl	Cl	OH	H	$CH(CH_3)_2$	$2{\cdot}0\times10^{-6}$	55
Phenethylamine*	H	H	H	H	H	$1{\cdot}1\times10^{-6}$	100
Phenelzine	H	H	H	H	NH_2	$3{\cdot}8\times10^{-6}$	29
Tranylcypromine	H	H	—CH_2—		H	$1{\cdot}3\times10^{-6}$	85

* Unsubstituted reference molecule.

are influenced by the uptake of these substances into sympathetic nerves. Since the D-enantiomers are inactivated by uptake more slowly than the L-enantiomers, the effects of the D-enantiomers are potentiated less by uptake inhibition than those of the L-enantiomers.

4. Other inhibitors of catecholamine uptake

Several other drugs have been reported to inhibit the uptake of catecholamines into sympathetic nerves. Most of these compounds have the expected property of potentiating the actions of catecholamines on smooth muscle preparations and potentiating the effects of noradrenaline released by sympathetic nerve stimulation. Imipramine and desipramine (*N*-desmethylimipramine) are particularly potent in this respect (Sigg, Soffer & Gyermek, 1963; Thoenen *et al.* 1964*c*) and are also potent inhibitors of catecholamine uptake (Axelrod, Whitby & Hertting, 1961; Axelrod, Hertting & Potter, 1962; Hertting *et al.* 1961; Dengler *et al.* 1961; Titus & Spiegel, 1962; Iversen, 1965*a*). Desipramine is the most potent inhibitor of catecholamine uptake so far described; in the isolated rat heart a concentration of $1{\cdot}3 \times 10^{-8}$ M is sufficient to produce a 50% inhibition of noradrenaline uptake (Table 8.1). This drug also produces a marked potentiation of the effects of catecholamines after very small doses (0·1 mg/kg) *in vivo*. Other drugs which produce potentiations of the effects of catecholamines have also been shown to inhibit noradrenaline uptake. For instance amitryptyline and its derivative protryptyline inhibit the uptake of metaraminol in the mouse heart (Carlsson & Waldeck, 1965). The antihistamine drugs chlorpheniramine and tripelennamine potentiate the actions of exogenous norepinephrine in isolated rat atria and also inhibit the uptake of H^3-noradrenaline into the rat heart *in vivo* (Isaac & Goth, 1965). The correlation between the ability to inhibit catecholamine uptake and the ability to potentiate the actions of catecholamines seems so well established that it may be predicted that other drugs which potentiate the actions of catecholamines, such as methyl phenidate (Maxwell *et al.* 1959), will prove to be inhibitors of uptake.

Bretylium, guanethidine and chlorpromazine also act as uptake inhibitors (Hertting, Axelrod & Patrick, 1962; Dengler *et al.* 1961; Lindmar & Muscholl, 1964; Iversen, 1965*a*). In these cases the situation is complicated by the other pharmacological actions of the drugs. Guanethidine is able to potentiate the actions of catecholamines on isolated atria (Stafford, 1963), bretylium is so weak an inhibitor of uptake that a potentiation cannot be demonstrated, and chlorpromazine may not potentiate because of its adrenergic receptor blocking activity. Guanethidine, but not bretylium or chlorpromazine, produces a long-lasting depletion of noradrenaline stores (Cass & Spriggs, 1961). The major pharmacological actions of these drugs are probably not directly related to their ability to inhibit catecholamine uptake.

The short-acting monoamine oxidase inhibitors harmaline and harmine were found to inhibit the uptake of noradrenaline in the isolated perfused rat heart (Lindmar & Muscholl, 1964; Iversen, 1965*a*). In the same preparation the monoamine oxidase inhibitors phenelzine and tranylcypromine were also potent inhibitors of noradrenaline uptake (Iversen, 1965*a*) (Table 8.1). The unexpected combination of monoamine oxidase inhibition and uptake inhibition may have pharmacological consequences of importance in evaluating the actions and clinical usefulness of such drugs.

The diversity of structure and ancillary actions in the compounds known to act as inhibitors of catecholamine uptake emphasizes again the importance of considering multiple sites of action in drugs which act on adrenergic mechanisms. The structural specificity of uptake inhibitors is quite distinct from the structural specificities of adrenergic receptors or metabolizing enzymes. However, drugs which interact with one site may well have the ability to interact with another.

5. Inhibition of Uptake$_2$ by drugs

Many of the drugs tested as inhibitors of the uptake of noradrenaline at low perfusion concentrations in the isolated rat heart (Uptake$_1$) were also investigated on Uptake$_2$ (Iversen, 1965*b*; Burgen & Iversen, 1965). The experimental procedure used was similar to that previously described. Drugs were added to the perfusion medium which in this case contained a high concentration (5 μg/ml) of H^3-noradrenaline. The uptake of noradrenaline was then measured in the presence of various test concentrations of the drug.

The results of this study are summarized in Table 8.10. Among the sympathomimetic amines tested the results showed very striking differences in the drug sensitivities of the two catecholamine uptake processes in the rat heart. In almost every respect the structural requirements for inhibition of $Uptake_2$ differed from those of $Uptake_1$. In the case of $Uptake_2$ both *N*-substitution and β-hydroxylation increased inhibitory potency, whereas phenolic hydroxyl groups or α-methylation decreased affinity. In *N*-substituted compounds the *N*-methyl derivative appeared to be the optimum structure. Studies of the uptake of *N*-substituted amines by $Uptake_2$ have shown that both adrenaline and isoprenaline are accumulated in the rat heart by this process at rates exceeding those for comparable concentrations of noradrenaline (Iversen, 1965*b*; Callingham & Burgen, 1966).

The most striking difference between $Uptake_1$ and $Uptake_2$, however, was in the effects of *O*-methylation. This greatly reduced affinity for $Uptake_1$, particularly when the methoxy groups were on the *meta*-position. For $Uptake_2$, however, *O*-methylation greatly *increased* affinity; normetanephrine, for example, had an affinity sixty times higher than that of noradrenaline. Metanephrine proved to be the most potent inhibitor of $Uptake_2$ among the various compounds tested.

Cocaine, desipramine and phenoxybenzamine all had some activity as inhibitors of $Uptake_2$, but cocaine and desipramine were less effective as inhibitors of $Uptake_2$ than they were as inhibitors of $Uptake_1$. The clear differences in the drug sensitivities of the two uptake processes lends strong support to the conclusion that these are indeed distinct processes, mediated by different mechanisms. The high affinities of the naturally occurring metabolites, normetanephrine and metanephrine, for $Uptake_2$ are very interesting, although the physiological significance of this finding remains obscure. The finding that the *O*-methylated catecholamines were potent inhibitors of $Uptake_2$ while these compounds have only very weak effects on $Uptake_1$, together with the converse findings for L-metaraminol may offer a useful method for distinguishing between the two uptake processes in other tissues or *in vivo*.

Since the role of $Uptake_2$ in adrenergic transmission remains unknown it would be of little value to speculate on the possible effects of drug inhibition of this process.

Table 8.10. *Inhibition of noradrenaline Uptake$_2$ by sympathomimetic amines*

(From Burgen & Iversen, 1965.)

Drug	ID 50 for Uptake$_2$ (M×)	Relative affinity for Uptake$_2$	Ratio $\frac{\text{ID 50 Uptake}_2}{\text{ID 50 Uptake}_1}$
(±)-Metanephrine	$2{\cdot}9\times10^{-6}$	2585	0·068
(±)-Normetanephrine	$4{\cdot}2\times10^{-6}$	1785	0·021
(±)-Nylidrin	$7{\cdot}6\times10^{-6}$	985	9·1
(±)-Synephrine	$1{\cdot}2\times10^{-5}$	625	1·0
3,4-Dimethoxyphenethylamine	$3{\cdot}2\times10^{-5}$	234	0·16
N-Dimethyl-*p*-tyramine	$4{\cdot}6\times10^{-5}$	163	18·4
(±)-Adrenaline*	$5{\cdot}2\times10^{-5}$	144	37·1
Phenethylamine	$7{\cdot}5\times10^{-5}$	100	68·2
m-Tyramine	$9{\cdot}5\times10^{-5}$	79	187·0
Tyramine	$1{\cdot}0\times10^{-4}$	75	223·0
(±)-Amphetamine	$1{\cdot}1\times10^{-4}$	68	239·0
3,4,5-Trimethoxy-phenethylamine	$1{\cdot}2\times10^{-4}$	62	0·008
(±)-Isoprenaline	$1{\cdot}2\times10^{-4}$	62	4·8
(±)-*N*-Ethylnoradrenaline	$1{\cdot}8\times10^{-4}$	42	19·6
p-Methoxyphenethylamine	$2{\cdot}0\times10^{-4}$	37	20·0
(±)-Noradrenaline*	$2{\cdot}5\times10^{-4}$	30	374·0
(±)-Methoxamine	$2{\cdot}7\times10^{-4}$	28	0·27
Dopamine	$4{\cdot}0\times10^{-4}$	19	2350·0
(−)-Metaraminol	$> 5{\cdot}0\times10^{-4}$	< 15	> 6600·0

* Affinities determined by direct analysis of uptake kinetics (Iversen, 1965*b*).

6. Effects of drugs on catecholamine uptake in isolated particles

There has been considerable interest in recent years in studies of the actions of drugs on isolated amine storage particles from the adrenal medulla or from adrenergic nerves. Such studies have the considerable advantage that the system under investigation is relatively simple. However, it is clear that extreme caution must be employed in extrapolating the results of such investigations to explain drug effects in intact tissues. The storage particles in adrenergic nerve endings represent only one part of the complex functional unit involved in noradrenaline synthesis, storage, metabolism, uptake and release. Furthermore, many of the drug studies have been performed on isolated amine storage particles from the adrenal medulla. As discussed in the previous chapter, such particles are not strictly comparable to those found in adrenergic nerves, and extrapolation of drug effects on one system to explain drug actions on the other is unjustified.

Isolated amine storage particles from both adrenal medulla (Weil-Malherbe & Posner, 1963; McLean & Cohen, 1963) and from adrenergic nerves (Euler & Lishajko, 1961, 1963; Euler, Stjärne & Lishajko, 1963; Stjärne, 1964) spontaneously release catecholamines if incubated in isotonic sucrose or potassium chloride. This release is faster at 37 °C than at 4 °C. The presence of low concentrations of reserpine (10^{-5}–10^{-6} M) inhibits this spontaneous release in both particles (Weil-Malherbe & Posner, 1963); Stjärne (1964), however, found reserpine at these concentrations to have no effect on the spontaneous release of catecholamines from medullary particles. At higher concentrations (10^{-4} M) reserpine promotes the release of the catecholamine content of storage particles. At low concentrations reserpine also inhibits the uptake of exogenous catecholamines into both types of storage particles (Kirshner, 1962; Carlsson, Hillarp & Waldeck, 1963; Mirkin, Giarman & Friedman, 1963; Euler & Lishajko, 1964; Euler, Stjärne & Lishajko, 1963, 1964; Jonasson, Rosengren & Waldeck, 1964; Stjärne, 1964; Potter & Axelrod, 1963). Stjärne (1964) found that reserpine produced a 50 % inhibition of noradrenaline uptake into medullary or splenic nerve particles at concentrations between 2×10^{-7} and 7×10^{-7} M, indicating that the drug acts potently on catecholamine uptake in isolated particle preparations (Fig. 8.2). It was also found (Stjärne, 1964) that one reserpine molecule could effectively inhibit the uptake of several hundred molecules of catecholamine, indicating that the drug probably interferes with some specific uptake mechanism in the particles, rather than simply competing with catecholamines for intraparticular storage sites.

Since both the uptake and spontaneous release of catecholamines in isolated storage particles are temperature-dependent and can be inhibited by reserpine it seems that not only amine uptake but also amine release is mediated by some specific chemical mechanism. The properties of the particle systems seem to rule out the possibility of a simple 'pump and leak' process, where the concentration of catecholamine inside the particles would be determined by the balance between the opposing processes of active transport of amines into the particles and passive diffusional leakage outwards. Nevertheless, the particle catecholamines, at least in adrenergic nerve granules, are in rapid equilibrium with extraparticulate catecholamines. When adrenergic nerve particles are incubated with radioactive noradrenaline the labelled amine rapidly exchanges with the endogenous noradrenaline in the particles (Stjärne, 1964): as much as 75 % of the particle content of noradrenaline can exchange in this way.

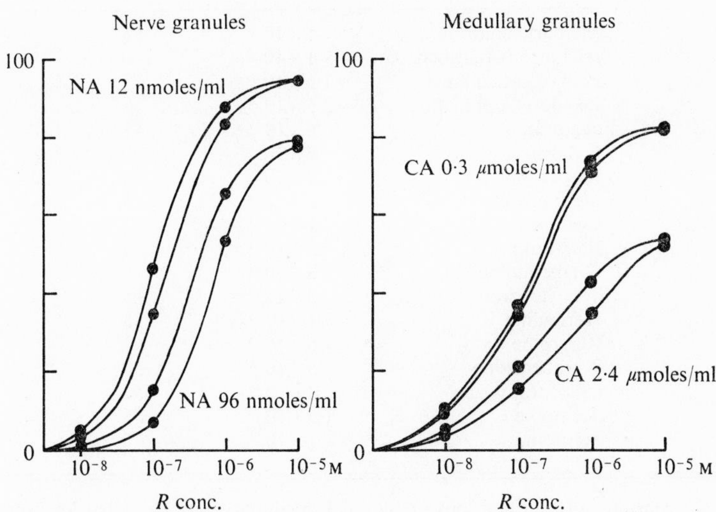

Fig. 8.2. Inhibitory effects of reserpine (*R*) on uptake of radioactive noradrenaline into bovine splenic nerve granules and adrenal medullary granules CA = catecholamine. Particle preparations were incubated for 15 min at 37 °C with two different concentrations of DL-noradrenaline-H^3, ATP and Mg^{2+} and various concentrations of reserpine. Results of two series of experiments are shown. (From Stjärne, 1964.)

In adrenergic nerve particles low concentrations of phenoxybenzamine (4×10^{-4} M) or segontin (3×10^{-5} M to 3×10^{-6} M) acted like reserpine in inhibiting the spontaneous release of noradrenaline (Stjärne, 1964). These drugs, however, had the opposite effect in medullary particle preparations, in which they accelerated the spontaneous release of catecholamines. These results again emphasize the existence of qualitative differences in the properties of the two types of amine storage particle.

Table 8.11. *Effect of various drugs on the uptake of catecholamines into bovine adrenal medullary storage granules* in vitro

(From Carlsson, Hillarp & Waldeck, 1963.)

Drug	Drug concentration (M ×)	Percentage inhibition of catecholamine uptake
Mercuric chloride	5×10^{-6}	95
p-Chloromercuribenzoate	1×10^{-3}	100
Phenoxybenzamine	$1{\cdot}6 \times 10^{-5}$	55
Dichloroisoprenaline	$3{\cdot}5 \times 10^{-4}$	50
Segontin	6×10^{-7}	45
Nylidrin	2×10^{-5}	50
Imipramine	3×10^{-5}	50
Desipramine	3×10^{-5}	25
Cocaine	1×10^{-3}	40
Bretylium	5×10^{-4}	10
Guanethidine	8×10^{-4}	15
Tranylcypromine	5×10^{-4}	65
Harmaline	1×10^{-5}	55
Reserpine	8×10^{-8}	63
Tetrabenazine	5×10^{-6}	50
Chlorpromazine	2×10^{-5}	50
Tyramine	$1{\cdot}5 \times 10^{-4}$	55
Serotonin	1×10^{-4}	45

Suspensions of purified bovine adrenal medullary storage particles were incubated for 30 min at 31 °C with C^{14}-labelled adrenaline or noradrenaline at a concentration of approximately 3×10^{-4} M. Incubation medium also contained 0·31 M glycyl glycine buffer pH 7·3 and $2{\cdot}5 \times 10^{-3}$ M ATP and $MgCl_2$. Results are mean values for two or more experiments.

In medullary particles high concentrations of tyramine and other related amines accelerate the spontaneous release of catecholamines (Schümann & Wegmann, 1960; Schümann & Philippu, 1962); at the same time a stoichiometric replacement of catecholamines by tyramine in the particle stores was observed. The following amines were active in displacing catecholamines: tyramine, β-phenylethylamine, amphetamine, methamphetamine, and ephedrine. Catecholamine release was not promoted by the following: dopamine, acetylcholine, nicotine, histamine or cocaine. Cocaine also failed to influence the release of catecholamines by tyramine.

Many drugs have been tested as inhibitors of catecholamine uptake in isolated medullary particles (Table 8.11) (Kirshner, 1962; Carlsson *et al.* 1963). The uptake process in such particles does not have the same spectrum of drug sensitivity as that described for catecholamine uptake in intact tissues. The adrenergic blocking agents are effective in both intact tissues and in medullary particles as inhibitors of catecholamine uptake. In the medullary particles, however, the potent inhibitors of the tissue uptake of noradrenaline, cocaine, imipramine and chlorpromazine were relatively ineffective. The differences in the drug sensitivities of catecholamine uptake in isolated particles and in intact tissues support the view that the uptake of catecholamines into storage particles cannot explain the ability of adrenergic nerves to accumulate exogenous catecholamines (chapter 7). It is interesting to note that the drug sensitivity of the uptake process in medullary storage particles more closely resembles that described for $Uptake_2$ in the rat heart than it resembles $Uptake_1$. Thus both $Uptake_2$ and the particle uptake system are inhibited by phenoxybenzamine and tyramine at 10^{-4}–10^{-5} M, while being only weakly inhibited by similar concentrations of cocaine, desipramine or metaraminol. Both processes also lack stereochemical specificity for the isomers of noradrenaline and adrenaline.

B. DRUGS WHICH AFFECT THE STORAGE OF NORADRENALINE

Since the discovery that reserpine treatment produced a long-lasting depletion of catecholamines from the brain and other tissues of experimental animals (Holzbauer & Vogt, 1956; Carlsson, Rosengren, Bertler & Nilsson, 1957) there have been many reports of drug-induced changes in the storage levels of catecholamines. However, as discussed above, changes in the storage level of noradrenaline may be the result of many different types of drug effect. In most cases the exact mechanism of action of the various drugs which influence the storage levels of noradrenaline remains unknown. The literature in this field is now so extensive that only a small part can be discussed in this chapter. The various examples discussed here illustrate the diversity of pharmacological effects which can give rise to changes in the levels of noradrenaline in tissue stores, and the sparsity of knowledge of the detailed mechanism of action of such drugs.

1. Reserpine

The release of catecholamines from peripheral tissues by reserpine has been extensively studied (see review by Shore, 1962). Small doses of reserpine (0·1 mg/kg) are sufficient to produce an almost complete depletion of the noradrenaline content of tissues such as heart and spleen. The depletion occurs only slowly after the administration of reserpine and is complete within 12–24 h. The tissue amine content then remains severely depleted for several days and does not return to normal levels until 7–14 days after the original treatment (Carlsson *et al.* 1957). In reserpine-treated tissues adrenergic transmission is blocked and the tissues no longer respond to indirectly acting sympathomimetic amines such as tyramine.

The depletion of noradrenaline stores by reserpine is not due to a release of noradrenaline from the adrenergic nerves. The studies of Kopin & Gordon (1962, 1963) showed that in the intact animal the administration of reserpine produced a release not of noradrenaline but mainly of deaminated metabolites of the catecholamine into the circulation (Table 8.12). In the isolated hearts of rats pretreated with reserpine, H^3-noradrenaline rapidly disappeared from the tissue when infusion of the labelled amine was stopped; again the amine was not released as unchanged noradrenaline but largely as deaminated metabolites (Kopin, Hertting & Gordon, 1962). Similar results were obtained by Chang, Costa & Brodie (1964).

In reserpinized tissues exogenous noradrenaline or adrenaline do not accumulate when the tissues are exposed to the catecholamines (Muscholl, 1960; Axelrod *et al.* 1961*b*; Hertting *et al.* 1961). Dengler *et al.* (1961) also reported that reserpine inhibited the uptake of tritiated noradrenaline into tissue slices incubated *in vitro*. These apparently unequivocal demonstrations of the ability of reserpine to inhibit catecholamine uptake have been widely accepted, and theories have been elaborated which attempt to explain the actions of reserpine on catecholamine storage on this basis (Brodie & Beaven, 1963). However, in view of recent evidence it seems unlikely that this view is correct. After reserpine treatment the main effect of the drug appears to be on the processes involved in the intraneuronal storage of noradrenaline. In reserpinized tissues the uptake of catecholamines can

Table 8.12. *Effects of drugs on the urinary excretion of H^3-noradrenaline and its metabolites*

(From Kopin & Gordon, 1963.)

Drug and dose	Percentage change in urinary excretion of H^3-noradrenaline and its metabolites: Total H^3	Noradrenaline	Normetanephrine	Deaminated catechols
Tyramine (10 mg/kg)	+36	+284	+245	0
Reserpine (2·5 mg/kg)	+89	+ 46	+ 8	+330
Pheniprazine (10 mg/kg)	−25	—	—	—
Pheniprazine + reserpine	+40	+164	+145	0
Guanethidine (15 mg/kg)	+66	+ 73	+ 35	+318
α-Methyl DOPA (50 mg/kg)	+60	+102	+ 13	+392
DMPP (0·1 mg/kg, 12 doses)	+31	+ 37	+ 37	− 8
Atropine (5 mg/kg) and acetylcholine (0·012 mg/kg, 12 doses)	+ 9·3	+ 23	+ 26	+ 5

Rats received an intravenous administration of 100 μc DL-H^3-noradrenaline. Urine was collected during the period from 10–14 h after the administration of H^3-noradrenaline and analysed for tritiated metabolites. During this time the urinary metabolites derive largely from H^3-noradrenaline released from storage sites in tissues. Drugs were administered before or at the beginning of the urine collection period.

occur at a normal rate, but since the accumulated catecholamines can no longer be retained at intraneuronal storage sites they are rapidly metabolized and leave the tissue. Thus, no accumulation of exogenous catecholamine can be observed; experiments in which the tissue accumulation of adrenaline or noradrenaline was measured several minutes or even hours after the initial uptake have thus generally yielded negative results.

Direct evidence for this hypothesis is available. Kopin *et al.* (1962) found that the initial uptake of tritiated noradrenaline in the isolated hearts from reserpinized rats did not differ from that in normal hearts, although the subsequent retention of the accumulated amine was markedly impaired. Lindmar & Muscholl (1964) infused a fixed amount (500 ng) of noradrenaline through the isolated rat heart during a 10 min period. The perfusate outflow was collected and analysed for unchanged noradrenaline; the amount of noradrenaline taken up during passage through the tissue could thus be estimated by the difference between the input and outflow concentrations of noradrenaline. An analysis of the noradrenaline content of the heart at the end of the perfusion showed that in normal hearts the noradrenaline taken up from the perfusing medium could be accounted for almost entirely as unchanged noradrenaline which accumulated in the tissue. In reserpinized hearts, however, although the amount of noradrenaline taken up from the perfusing medium was the same as in normal hearts, the amine failed to accumulate in the tissue. We were able to repeat this experiment in the isolated rat heart and by using H^3-noradrenaline to show that the 'missing' noradrenaline in the reserpinized hearts could be accounted for almost entirely by tritiated deaminated metabolites which were formed in the tissue and which were released into the perfusing medium (Table 8.13) (Iversen, Glowinski & Axelrod, 1965). Gillespie & Kirpekar (1965) have also reported that the amount of noradrenaline removed during passage of blood through the cat spleen was not reduced in reserpinized animals.

Further direct evidence for an unimpaired ability to take up exogenous catecholamines in reserpinized tissues was provided by the histochemical studies of Hamberger *et al.* (1964). In the rat iris an intense fluorescence accumulated in the adrenergic ground plexus after administration of noradrenaline or α-methyl noradrenaline to reserpinized animals. The only difference between the appearance of the iris after noradrenaline administration in normal or reserpinized animals was that the accumulated amines in the reserpinized tissue were diffusely distributed throughout the sympathetic preterminal axons, instead of being concentrated in the terminal varicosities.

Further evidence which supports the view that catecholamine uptake is unimpaired in reserpinized tissues is provided by the numerous experiments which have demonstrated a 'refilling' of the noradrenaline stores in such tissues on exposure to noradrenaline or related amines. The ability of reserpinized tissues to respond to sympathetic nerve stimulation or to tyramine can be restored temporarily by such treatment (Burn & Rand, 1958, 1960; Rosell & Sedvall, 1961; Muscholl, 1960; Crout Muskus & Trendelenburg, 1962; Fawaz & Simaan, 1963; Trendelenburg & Pfeffer, 1964). If reserpine did in fact inhibit noradrenaline uptake the tissues would be expected to exhibit

Table 8.13. *The fate of* H^3*-noradrenaline in the isolated hearts of normal and reserpinized rats*

(From Iversen, Glowinski & Axelrod, 1965.)

	Normal	Reserpinized
Total uptake of H^3-NA during 10 min perfusion (mμg/g)	50·3±8·4	52·0±7·4
	Percentage of total H^3-NA uptake	
	Normal	Reserpinized
1. Remaining in tissue		
(*a*) Unchanged H^3-NA	95·0±5·5	48·0±2·5
(*b*) Deaminated catechol metabolites	<1·0	6·0±0·4
(*c*) Normetanephrine	2·2±0·2	2·2±0·2
(*d*) *O*-Methylated deaminated metabolites	Nil	*c*. 3·0
Total in tissue	98·0±5·8	61·0±3·6
2. Returned to medium as metabolites		
(*a*) Deaminated catechol metabolites	< 1·0	18·0±1·0
(*b*) Normetanephrine	< 1·0	< 1·0
(*c*) *O*-Methylated deaminated metabolites	< 2·0	*c*. 18·0
Total returned to medium	1·7±3·9	36·0±6·5

Isolated hearts were perfused with a medium containing 3·38 mμg/ml H^3-noradrenaline (NA) for 10 min. At the end of this perfusion the tissue and the perfusate outflow were analysed for unchanged H^3-NA and metabolites. Each value is the mean for twelve normal hearts and for twelve reserpinized hearts (24 h after 5 mg/kg i.p.)±S.E.M. Deaminated catechols represent 3,4-dihydroxymandelic acid and the corresponding glycol. *O*-Methylated metabolites are 3-methoxy-4-hydroxymandelic acid and the glycol.

an immediate supersensitivity to exogenous catecholamines similar to that seen after cocaine treatment. However, the supersensitivity that develops only slowly in reserpinized tissues does not resemble that seen after cocaine (Trendelenburg, 1963; Fleming & Trendelenburg, 1961). Furthermore, in reserpinized tissues in which supersensitivity has not yet developed, drugs such as tyramine and cocaine which inhibit noradrenaline uptake can still produce a normal potentiation of the pharmacological effects of noradrenaline (Furchgott *et al.* 1963; Lindmar & Muscholl, 1965).

The conclusion that reserpine does not interfere with the uptake of catecholamines into sympathetic nerves thus seems to be well established. The actions of the drug appear to be restricted to intraneuronal catecholamine storage mechanisms. The evidence discussed earlier in this chapter, that reserpine potently inhibits both the uptake and release of catecholamines in adrenergic nerve storage particles, strongly suggests that these represent the site of action of the drug. The precise mechanism involved in the catecholamine depletion produced by the drug, however, remains obscure. It has been suggested that an inhibition of dopamine uptake into neuronal storage granules in reserpinized tissues may result in a blockade of noradrenaline biosynthesis (Kirshner, Rorie & Kamin, 1963). However, there is evidence that catecholamine synthesis is unimpaired in reserpinized tissues, although the ability to retain the product, noradrenaline, is lacking (Bertler, 1961; Glowinski, Iversen & Axelrod, 1966). Furthermore, recent evidence suggests that the synthesis of noradrenaline from DOPA takes place entirely within the storage particles (chapter 5), so that the entry of exogenous dopamine into these particles is not a requisite for normal synthesis. The action of reserpine on isolated storage particles does not seem to involve the entry of catecholamines across the particle membrane, which is apparently freely permeable to catecholamines (Stjärne, 1964). It has been suggested that reserpine acts to block some specific mechanism inside the storage particle by which catecholamines enter or leave the storage sites (Stjärne, 1964). The action of reserpine could thus be postulated as follows. Reserpine prevents the entry of catecholamines into a storage compartment within the adrenergic nerve storage particles. Noradrenaline is normally synthesized in the storage particles but the synthetic machinery is located outside the storage compartment, possibly on the particle membrane; newly synthesized noradrenaline is thus unable to enter the storage compartment and is consequently lost from the particles by passive diffusional leakage. The noradrenaline which leaks out of the particles is rapidly destroyed by monoamine oxidase within the axoplasm of the nerve fibre. As the noradrenaline content of the storage sites is lost by neuronal release and diffusion it cannot be replaced, so that the noradrenaline content rapidly falls. In keeping with this view the rate of depletion of noradrenaline in reserpinized tissues is dependent on the activity of the sympathetic nerve supply (Kärki, Paasonen & Vanhakartano, 1959; Benmiloud & Euler, 1963; Weiner, Perkins &

Sidman, 1962). The noradrenaline released from reserpinized tissues by nerve impulses can be recaptured by the normal process of re-uptake, but the catecholamine cannot then be returned to the normal storage sites and is consequently rapidly destroyed by monoamine oxidase. This scheme is admittedly speculative and is not intended to be a consensus of current opinions about the mechanism of action of reserpine which have always been controversial and will probably continue to be so (see Costa *et al.* 1966).

2. The actions of sympathomimetic amines

(*a*) *Classification of sympathomimetic amines*

Barger & Dale (1910) demonstrated that a wide variety of amines structurally related to adrenaline and noradrenaline have biological actions similar to those of adrenaline and noradrenaline. Tainter & Chang (1927) and Tainter (1929) showed that the actions of two of these amines, tyramine and ephedrine, were antagonized by doses of cocaine which caused supersensitivity to the actions of adrenaline (Fröhlich & Loewi, 1910). This finding has been aptly termed 'the cocaine paradox'. Burn & Tainter (1931) confirmed the differential action of cocaine on adrenaline and tyramine effects on isolated tissues, and also showed that a similar difference was obtained in tissues which had been sympathetically denervated. In the denervated nictitating membrane, they found that the actions of adrenaline were greatly potentiated, whilst the actions of tyramine were abolished. Burn (1932) confirmed this observation by showing that the vasoconstrictor effects of tyramine and ephedrine were abolished in the denervated cat's foreleg, and made the first suggestion that the effects of these amines might depend on the integrity of adrenergic nerve endings and that they might not act directly on adrenergic receptors.

Fleckenstein & Bass (1953) studied the actions of a large number of sympathomimetic amines in the presence or absence of cocaine, and Fleckenstein & Burn (1953) performed a similar extensive study of the actions of sympathomimetic amines on denervated tissues. These studies were further extended by Fleckenstein & Stöckle (1955). It was concluded that three groups of amines could be distinguished: (1) those which are potentiated by denervation and cocaine—'direct'-acting amines; (2) those which are not much affected by either procedure—'mixed'-acting amines; and (3) those which are clearly less effective after denervation or cocaine—'indirect'-acting amines. The prototypes for these three classes were noradrenaline (direct), ephedrine (mixed) and tyramine (indirect). Similar studies were conducted by Innes & Kosterlitz (1954) and by Holtz *et al.* (1960).

More recently reserpine treatment has been used as an additional tool to distinguish between the three classes of sympathomimetic amines. Carlsson *et al.* (1957) showed that tyramine was without activity in reserpine-treated cats. Burn & Rand (1958) confirmed and extended this finding. They showed that in reserpine-treated animals the peripheral stores of noradrenaline were abolished, and that under these conditions the actions of indirectly acting amines were abolished or diminished, but could be temporarily restored by infusions of noradrenaline (Burn & Rand, 1960). Burn & Rand (1958) put forward the now widely accepted hypothesis that tyramine and other indirectly acting amines produce their sympathomimetic effects by releasing noradrenaline from tissue storage sites where it is normally bound and prevented from activating the tissue receptors. A wide range of sympathomimetic amines have been tested in reserpinized preparations (Innes & Krayer, 1958; Fleming & Trendelenburg, 1961; Fleming & Schmidt, 1962; Trendelenburg, Muskus, Fleming & Gomez, 1962*a*, *b*; Schmidt & Fleming, 1963).

On the basis of these and previous results concerning the actions of sympathomimetic amines on denervated and cocaine treated tissues, it has been possible to make detailed classifications of the amines into the three groups originally described. The subject has recently been reviewed by Trendelenburg (1963) and the following classification is based on Table 1 of Trendelenburg's review. It should be emphasized that a clear-cut distinction between direct, mixed and indirect acting amines is not possible, but that the amines possess a continuous spectrum of actions ranging from almost pure direct to almost pure indirect, and that each substance probably has some direct actions on adrenergic receptors and some indirect actions on noradrenaline stores, so that the overall action of each amine will depend on the relative potency of its two types of action. Since these may vary in different tissues, and on different adrenergic receptors, amines may be classified as indirect acting in one tissue, though they may have mixed or direct actions on another tissue of the same animal. However, the following tentative classification can be made:

Direct-acting amines	Mixed-acting amines	Indirect-acting amines
Noradrenaline	Norsynephrine	*p*-Tyramine
Adrenaline	Phenylethanolamine	β-Phenylethylamine
Cobefrin	Phenylpropanolamine	Amphetamine
Dihydroxyephedrine	Metaraminol	Paredrine
Dopamine	Ephedrine	Methamphetamine
Epinine	*m*-Tyramine	Mephentermine
Isoprenaline		Pholedrine
Phenylephrine		
Synephrine		
Norphenylephrine		

The structural requirements for direct and indirect actions seem to be fairly consistent: direct-acting amines comprise all the catechol amines (with or without a β-hydroxyl group) and certain monophenols which have a β-hydroxyl group and a phenolic hydroxyl in the meta-position (e.g. phenylephrine). Indirect-acting amines include non-phenolic compounds, or monophenols without a β-hydroxyl group. It is clear that these distinctions cannot be made with any rigidity, owing to the graded transition from direct- to indirect-acting amines.

(*b*) *The evidence in favour of the noradrenaline displacement theory*

Current opinion holds that tyramine and similar amines exert their sympathomimetic effects through the release of noradrenaline from storage sites in peripheral tissues. Direct stimulation of noradrenaline release from the adrenal medulla by tyramine does not seem to be of any importance, since most workers have failed to detect an increase in the plasmal levels of catecholamines in adrenal venous blood after injections of tyramine (e.g. Stjärne, 1961).

Direct evidence of the displacement of NA from peripheral tissues has been of two types. First, it has been shown that tyramine and other indirectly acting amines promote an increased overflow of noradrenaline in the venous outflow or perfusate effluent of various organs. This has been demonstrated for tyramine in the cat spleen (Stjärne, 1961), for β-phenylethylamine and tyramine in the isolated perfused rabbit heart (Lindmar & Muscholl, 1961), for tyramine, β-phenylethylamine, amphetamine, ephedrine, paredrine and mephentermine in the dog heart (Chidsey, Harrison & Braunwald, 1962). Radioactively labelled noradrenaline was displaced by tyramine in isolated cat atria (Burn & Burn, 1961) and in dog heart (Mueller & Shideman, 1962).

Further evidence for the view that tyramine acts through the release of noradrenaline is provided by reports that the administration of large doses of tyramine *in vivo* causes a partial depletion of the noradrenaline stores in various tissues (Weiner *et al.* 1962; Porter *et al.* 1963).

A different line of evidence for the action of tyramine on noradrenaline stores is that the responsiveness to tyramine can be restored in reserpinized animals by a preliminary infusion of noradrenaline or some noradrenaline precursor (Burn & Rand, 1958, 1960; Crout, Muskus & Trendelenburg, 1962; Muscholl, 1960). These reports have suggested that an infusion of noradrenaline in reserpinized animals temporarily 'refills' the tissue stores, so that tyramine is then able to act in the normal way by displacing noradrenaline. However, these results are puzzling, since pretreatment with reserpine has repeatedly been shown to reduce greatly the ability of tissues to accumulate exogenous noradrenaline, yet the full responsiveness to tyramine can be restored even though less than 1 % of the normal noradrenaline content appears to have been 'refilled'. This aspect will be discussed more fully in chapter 9.

Thus the various lines of evidence seem to confirm the hypothesis that tyramine acts by displacing noradrenaline from peripheral storage sites (for dissenting viewpoint see Zaimis, 1964). No clear idea of the mechanism of this release, however, has yet been proposed. It seems to be tacitly assumed that tyramine displaces noradrenaline from binding sites in the tissues, and that this depends on a mass action effect, whereby tyramine replaces some noradrenaline molecules in the store in a stoichiometric ratio of one tyramine molecule replacing one noradrenaline molecule. It has further been suggested that tyramine has first to gain entry to the intracellular storage sites, and that this process may be mediated by a transfer mechanism which is sensitive to cocaine inhibition, which would thus explain the inhibitory effects of cocaine on the effects of tyramine (Furchgott *et al.* 1963). The sensitivity of the tyramine response to cocaine suggests that tyramine may be transported into the heart by the mechanism which normally accumulates noradrenaline and adrenaline. The findings that the tyramine response is also abolished by guanethidine and bretylium and that phenoxybenzamine (Swaine, 1963) can also prevent the release of noradrenaline by tyramine support this hypothesis, since these drugs are known to inhibit noradrenaline uptake. Tyramine is also known to act as an inhibitor of noradrenaline uptake, and in isolated guinea-pig atria the presence of tyramine prevented the 'refilling' of the noradrenaline stores of reserpinized animals (Trendelenburg, 1962). Axelrod, Gordon, Hertting, Kopin & Potter (1962) showed that C^{14}-tyramine was accumulated by the isolated rat heart. The tissue level of tyramine was found to rise to several times the normal endogenous noradrenaline content, indicating that the uptake of tyramine could not have been accompanied by a stoichiometric release of noradrenaline. Tyramine uptake and the subsequent displacement of endogenous noradrenaline from tissue stores thus appear to be distinct processes, and are probably mediated by different mechanisms.

The ability of tyramine and many other sympathomimetic amines to act as potent inhibitors of noradrenaline uptake can explain why the very small amounts of noradrenaline released by these amines have such potent effects, since the inhibition of noradrenaline re-uptake would potentiate the actions of the released noradrenaline. This may explain the results of Weiner, Draskóczy & Burack (1962) who found that in the adrenalectomized dog the amount of noradrenaline released into the circulation after tyramine was very small, but after an equipressor dose of nicotine there was a much larger release of noradrenaline into the circulation. Similarly, Lindmar & Muscholl (1961) observed that the amount of noradrenaline liberated from the perfused rabbit heart by tyramine was much less than the amount of noradrenaline released by dimethyl-phenyl-piperazinium iodide (DMPP) (a nicotine-like agent) although the two drugs produced equal cardiostimulant effects. Lindmar & Muscholl (1961, 1965) postulate that tyramine not only releases noradrenaline from the tissue stores but also enhances the effect of the released noradrenaline at the receptor sites.

The rapid conversion of tyramine and related amines to their β-hydroxylated derivatives inside sympathetic nerves may have important implications in explaining the mode of action of these amines. As shown by Musacchio, Kopin & Weise (1965), β-hydroxylated amines have a much higher affinity for intraneuronal particle storage sites than their non-β-hydroxylated analogues. It may be that tyramine and other such amines have to be converted to their β-hydroxylated derivatives before they are able to enter the particle storage sites and displace noradrenaline.

3. Actions of other drugs on noradrenaline storage

The examples described above illustrate noradrenaline depletion by two quite distinct types of action; in the case of reserpine, noradrenaline depletion results from an impairment of noradrenaline storage in intraneuronal particles followed by an uncontrolled metabolism of noradrenaline by monoamine oxidase. Sympathomimetic amines, on the other hand, release unchanged noradrenaline, perhaps by displacing the catecholamine from intraneuronal particles. The remaining examples, however, largely comprise drugs whose mode of action is not well understood.

Tetrabenazine and a group of related benzoquinolizines, like reserpine, produce a severe depletion of catecholamines from the brain (Pletscher, Brossi & Gey, 1962). Unlike reserpine, tetrabenazine is less effective in depleting the serotonin content of the brain and is without effect on peripheral stores of catecholamines or serotonin (Quinn, Shore & Brodie, 1959). The actions of the benzoquinolizines, although resembling to some extent those of reserpine, have not been studied in sufficient detail to allow any suggestions as to their mode of action.

The α-methyl amino acids α-methyl-*m*-tyrosine and α-methyl DOPA produce a long-lasting depletion of central and peripheral catecholamine stores. As discussed in chapters 5 and 6, this depletion of catecholamine is accompanied by an apparently stoichiometric replacement by the products metaraminol and an α-methyl noradrenaline formed from the amino acids *in vivo*. Although it seems well established that such a replacement does occur after the administration of the α-methyl amino acids, the mechanisms involved are not clear. If the α-methylated amines do replace noradrenaline as false neurochemical transmitters, it is difficult to understand the prolonged depletion of noradrenaline levels resulting from the administration of these drugs. One possible explanation for this apparently anomalous situation is the fact that both α-methyl noradrenaline and metaraminol have very high affinities for the catecholamine uptake process in sympathetic nerves (Table 8.3). In addition, the α-methylated amines cannot be destroyed within the nerve by monoamine oxidase, and they are bound with high affinities to the intraneuronal storage particles (Musacchio *et al.* 1965). Once in the storage sites in sympathetic nerves the α-methylated amines might therefore be expected to persist for long periods. After release by nerve impulses the α-methylated amines would be very efficiently recaptured by the catecholamine uptake process and would be returned to intraneuronal storage particles with little loss, since the amines cannot be metabolized within the nerve. Added to this is the possibility that the presence of the α-methylated amines within the nerve may inhibit the biosynthesis of noradrenaline.

Segontin produces a marked and prolonged depletion of noradrenaline from the brain and peripheral tissues of rats and rabbits (Schöne & Lindner, 1960; Jurio & Vogt, 1965; Mackenna, 1965). The noradrenaline content of rabbit heart rose to one-third normal levels after 24 h (Mackenna, 1965). The noradrenaline content of segontin-depleted hearts could be restored temporarily by injections of noradrenaline (Mackenna, 1965). It is interesting to note that segontin is a potent inhibitor of catecholamine uptake in isolated medullary (Carlsson *et al.* 1963) and splenic nerve particles (Euler *et al.* 1964). This might suggest that segontin depletes catecholamines by a mechanism similar to that described for reserpine. Segontin is also capable of inhibiting noradrenaline uptake into sympathetic nerves (Iversen, 1964).

The boron hydrides decaborane and pentaborane have been found to deplete brain and peripheral stores of catecholamines (Merritt, Schultz & Wykes, 1965). Noradrenaline levels can be restored temporarily to normal by injection of noradrenaline after decaborane depletion (Euler & Lishajko, 1965).

6-Hydroxydopamine can also produce a long-lasting and almost complete depletion of noradrenaline from the heart (Porter, Totaro & Stone, 1963). The mechanism of this effect is also obscure. Laverty, Sharman & Vogt (1965) were unable to detect 6-hydroxydopamine in the tissues of rats treated with this drug, although noradrenaline levels remained severely depleted for many days. Porter *et al.* (1965), however, injected radioactively labelled 6-hydroxydopamine and reported the presence of a radioactive compound in mouse hearts which was present in almost stoichiometric amounts to the displaced noradrenaline.

Guanethidine produces a long-lasting and complete depletion of peripheral noradrenaline stores (Cass, Kuntzman & Brodie, 1960; Butterfield & Richardson, 1961). Guanethidine, although similar in some respects to reserpine, releases noradrenaline not only in the form of deaminated metabolites but also to some extent as the free catecholamine (Kopin & Gordon, 1963) (Table 8.12). It has also been suggested on the basis of pharmacological evidence that reserpine and guanethidine have different sites of action (Muskus, 1964).

The aliphatic compound epsilon amino caproic acid produces a marked depletion of endogenous and exogenous noradrenaline from the mouse heart; other related aliphatic acids and amines were relatively inactive as noradrenaline depleting agents (Lippmann & Wishnik, 1965).

Certain drugs may cause a reduction in the levels of noradrenaline by indirect mechanisms, for example by producing an intense stimulation of peripheral adrenergic nerves or central sympathetic centres. Thus DMPP and other ganglion-stimulating agents can release nor adrenaline from peripheral stores (Kopin & Gordon, 1963) and agents such as morphine and β-tetrahydronaphthylamine can lower the concentration of noradrenaline in the hypothalamus (Laverty & Sharman, 1965).

4. Drug interactions

As discussed above, it is well known that the actions of tyramine and other indirectly acting sympathomimetic amines can be blocked by drugs such as cocaine, desipramine and other inhibitors of catecholamine uptake. This strongly suggests that the amines are taken up into sympathetic nerves by the catecholamine uptake process before they can have their effects on the noradrenaline stores. This finding may be of quite general application, in that the actions of many drugs which deplete noradrenaline stores can be antagonized by other drugs which inhibit the catecholamine uptake process.

Thus the actions of reserpine and guanethidine in producing a depletion of noradrenaline levels in peripheral tissues can be blocked by cocaine or bretylium (Callingham & Cass, 1962; Costa, Kuntzman, Gessa & Brodie, 1962). Stone *et al.* (1964) reported that the noradrenaline depletion produced by α-methyl-*m*-tyrosine, metaraminol, guanethidine and 6-hydroxydopamine could be prevented by the administration of a wide variety of drugs, including imipramine, desipramine, amitriptyline, protryptyline, desmethylamitryptyline, cocaine, tripelennamine, pipradol, methylphenidate, chlorpromazine and promazine—many of which are known to act as inhibitors of catecholamine uptake. Chlorpromazine can also prevent the depletion of brain catecholamines and serotonin produced by reserpine (Pletscher & Gey, 1961).

C. EFFECTS OF OTHER FACTORS ON NORADRENALINE UPTAKE AND STORAGE

Garattini & Valzelli (1958) and Sulser & Brodie (1960) reported that exposure of rats to high or low temperatures inhibited the release of brain serotonin by reserpine but the release of catecholamines is unaffected. Cold stress causes a marked increase in the urinary excretion of catecholamines and their metabolites and leads to a depletion of noradrenaline stores in peripheral tissues (Leduc, 1961). Stress can also produce changes in the uptake and retention of exogenous noradrenaline in tissues. Gillis (1965) reported that the uptake of H^3-noradrenaline in the rabbit heart was reduced if the animals were given i.p. injections of saline or mock injections before the injection of H^3-norepinephrine, suggesting that the stress induced by the handling had initiated a change in the mechanisms involved in the uptake and retention of noradrenaline. Gillis (1963) also reported that the uptake of H^3-noradrenaline in cat atria was increased if the cardioaccelerator nerve was stimulated immediately before exposure to the labelled noradrenaline. Little is known of the effects of sympathetic nerve activity on the noradrenaline uptake mechanism in sympathetic nerves. Blakeley & Brown (1964) suggest that the presence of impulses in the splenic nerves of the cat impairs the uptake of infused noradrenaline.

In newborn rats the ability of peripheral tissues such as the heart to take up and retain noradrenaline is almost totally lacking, the normal uptake and retention mechanisms develop rapidly between the 7th and

14th days after birth (Glowinski, Axelrod, Kopin & Wurtman, 1964). The hearts of 10-day-old rats are not able to accumulate as much noradrenaline per gram as adults, and the accumulated amine also disappears more rapidly than from the adult heart (Glowinski *et al.* 1964).

Hyperthyroid rats develop a supersensitivity to the effects of administered catecholamines. The hypertrophic hearts of these animals have a reduced capacity to take up and retain H^3-noradrenaline (Wurtman, Kopin & Axelrod, 1964).

Few studies have investigated the effects of environmental factors on the storage and metabolism of noradrenaline. In agreement with the well-documented effects of cold stress in increasing the rate of urinary excretion of catecholamines (Leduc, 1961), a recent report provides evidence for a more rapid turnover of noradrenaline in the peripheral sympathetic nerves of such animals (Oliverio & Stjärne, 1965). Other reports have claimed that the levels of noradrenaline in the heart and fat tissues of the rat are significantly increased by muscular excercise or by fasting (Nikkilä, Torsti & Pentillä, 1965); and that the levels of catecholamines in the retina of the frog, toad or rabbit are significantly higher in dark-adapted animals (Drujan, Diaz Borges & Alvarez, 1965).

REFERENCES

Axelrod, J. & Tomchick, R. (1960). Increased rate of metabolism of epinephrine and norepinephrine by sympathomimetic amines. *J. Pharmac. exp. Ther.* **130,** 367–369.

Axelrod, J., Hertting, G. & Patrick, R. W. (1961). Inhibition of H^3-norepinephrine release by monoamine oxidase inhibitors. *J. Pharmac. exp. Ther.* **134**, 325–328.

Axelrod, J., Whitby, L. G. & Hertting, G. (1961). Effect of psychotropic drugs on the uptake of H^3-norepinephrine by tissues. *Science* **133**, 383–384.

Axelrod, J., Hertting, G. & Potter, L. T. (1962). Effect of drugs on the uptake and release of H^3-norepinephrine in the rat heart. *Nature, Lond.* **194**, 297.

Axelrod, J., Gordon, E., Hertting, G., Kopin, I. J. & Potter, L. T. (1962). On the mechanism of tachyphylaxis to tyramine in the isolated rat heart. *Br. J. Pharmac. Chemother.* **19**, 56–63.

Barger, G. & Dale, H. H. (1910). Chemical structure and sympathomimetic actions of amines. *J. Physiol., Lond.* **41**, 19–59.

Benmiloud, M. & Euler, U. S. von (1963). Effects of bretylium, reserpine, guanethidine and sympathetic denervation on the noradrenaline content of the rat submaxillary gland. *Acta physiol. scand.* **59**, 62–66.

Bertler, A. (1961). Effect of reserpine on the storage of catecholamines in brain and other tissues. *Acta physiol. scand.* **51**, 75–83.

Blakeley, A. G. H. & Brown, G. L. (1964). The effect of nerve stimulation on the uptake of infused noradrenaline by the perfused spleen. *J. Physiol., Lond.* **172**, 19–20*P*.

Blakeley, A. G. H., Brown, G. L. & Ferry, C. B. (1963). Pharmacological experiments on the release of the sympathetic transmitter. *J. Physiol., Lond.* **167**, 505–514.

Blakeley, A. G. H., Brown, G. L. & Geffen, L. B. (1964). Uptake and re-use by the sympathetic nerves of the transmitter they release. *J. Physiol., Lond.* **173**, 22*P*.

Brodie, B. B. & Beaven, M. A. (1963). Neurochemical transducer systems. *Med. exp.* **8**, 320–351.

Brown, G. L. (1965). The Croonian Lecture 1964. The release and fate of the transmitter liberated by adrenergic nerves. *Proc. R. Soc.* B **162**, 1–19.

Brown, G. L. & Gillespie, J. S. (1957). The output of sympathetic transmitter from the spleen of the cat. *J. Physiol., Lond.* **138**, 81–102.

Brown, B. G. & Hey, P. (1956). Choline phenyl ethers as inhibitors of amine oxidase. *Br. J. Pharmac. Chemother.* **11**, 58–65.

Burgen, A. S. V. & Iversen, L. L. (1965). The inhibition of noradrenaline uptake by sympathomimetic amines in the rat isolated heart. *Br. J. Pharmac. Chemother.* **25**, 34–49.

Burn, G. P. & Burn, J. H. (1961). Uptake of labelled noradrenaline by isolated atria. *Br. J. Pharmac. Chemother.* **16**, 344–351.

Burn, J. H. (1932). The action of tyramine and ephedrine. *J. Pharmac. exp. Ther.* **46**, 75–95.

Burn, J. H. & Rand, M. J. (1958). The action of sympathomimetic amines in animals treated with reserpine. *J. Physiol., Lond.* **144**, 314–346.

Burn, J. H. & Rand, M. J. (1960). The effect of precursors of noradrenaline on the response to tyramine and sympathetic stimulation. *Br. J. Pharmac. Chemother.* **15**, 47–55.

Burn, J. H. & Tainter, M. L. (1931). An analysis of the effect of cocaine on the actions of adrenaline and tyramine. *J. Physiol., Lond.* **71**, 169–193.

Butterfield, J. L. & Richardson, J. A. (1961). Acute effects of guanethidine on myocardial contractility and catecholamine levels. *Proc. Soc. exp. Biol. Med.* **106**, 259–262.

Callingham, B. A. & Cass, R. (1962). The effects of bretylium and cocaine on noradrenaline depletion. *J. Pharm. Pharmac.* **14**, 385–389.

Callingham, B. A. & Burgen, A. S. V. (1966). The uptake of isoprenaline and noradrenaline by the perfused rat heart. *J. mol. Pharmac.* **2**, 37–42.

Cass, R. & Spriggs, T. L. B. (1961). Tissue amine levels and sympathetic blockade after guanethidine and bretylium. *Br. J. Pharmac. Chemother.* **17**, 442–450.

Cass, R., Kuntzman, R. & Brodie, B. B. (1960). Norepinephrine depletion as a possible mechanism of action of guanethidine (SU 5864), a new hypotensive agent. *Proc. Soc. exp. Biol. Med.* **103**, 871–872.

Carlsson, A. & Waldeck, B. (1965). Inhibition of H^3-metaraminol uptake by antidepressive and related agents. *J. Pharm. Pharmac.* **17**, 243–244.

Carlsson, A., Hillarp, N.-Å. & Waldeck, B. (1963). Analysis of the Mg^{++}–ATP dependent storage mechanism in the amine granules of the adrenal medulla. *Acta physiol. scand.* **59**, Suppl. 215.

Carlsson, A., Rosengren, E., Bertler, A. & Nilsson, J. (1957). Effect of reserpine on the metabolism of catecholamines. In *Psychotropic Drugs* (ed. Garrattini and Ghetti), pp. 363–372. Amsterdam: Elsevier.

Chang, C. C., Costa, E. & Brodie, B. B. (1964). Reserpine induced release of drugs from sympathetic nerve endings. *Life Sciences* **3**, 839–844.

Chidsey, C. A., Harrison, D. C. & Braunwald, E. (1962). Release of norepinephrine from the heart by vasoactive amines. *Proc. Soc. exp. Biol. Med.* **109**, 488–490.

Costa, E., Kuntzman, R., Gessa, G. L. & Brodie, B. B. (1962). Structural requirements for bretylium and guanethidine-like activity in a series of guanidine derivatives. *Life Sciences* **1**, 75–80.

Costa, E., Boullin, D. J., Hammer, W., Vogel, W. & Brodie, B. B. (1966). Interactions of drugs with adrenergic neurons. (Second International Catecholamine Symposium.) *Pharmacol. Rev.* **18**, 577–598.

Crout, J. R., Muskus, A. J. & Trendelenburg, U. (1962). Effect of tyramine on isolated guinea pig atria in relation to their noradrenaline stores. *Br. J. Pharmac. Chemother.* **18**, 600–611.

Dengler, H. J., Spiegel, H. E. & Titus, E. O. (1961). Effects of drugs on uptake of isotopic norepinephrine. *Nature, Lond.* **191**, 816–817.

Drujan, B. D., Diaz Borges, J. M. & Alvarez, N. (1965). Relationship between the contents of adrenaline, noradrenaline and dopamine in the retina and its adaptational state. *Life Sciences* **4**, 473–477.

Eakins, K. E. & Eakins, H. M. T. (1964). Adrenergic mechanisms and the outflow of aqueous humor from the rabbit eye. *J. Pharmac. exp. Ther.* **144**, 60–65.

Euler, U. S. von & Lishajko, F. (1961). Noradrenaline release from isolated nerve granules. *Acta physiol. scand.* **51**, 193–203.

Euler, U. S. von & Lishajko, F. (1963). Catecholamine release and uptake in isolated adrenergic nerve granules. *Acta physiol. scand.* **57**, 468–480.

Euler, U. S. von & Lishajko, F. (1964). Effect of reserpine on the uptake of catecholamines in adrenergic nerve granules. *Acta physiol. scand.* **60**, 217–222.

Euler, U. S. von & Lishajko, F. (1965). Uptake of catecholamines in the rabbit heart after depletion with decaborane. *Life Sciences* **4**, 969–972.

Euler, U. S. von, Stjärne, L. & Lishajko, F. (1963). Uptake of radioactively labelled DL-catecholamines in isolated adrenergic nerve granules with and without reserpine. *Life Sciences* **2**, 878–885.

Euler, U. S. von, Stjärne, L. & Lishajko, F. (1964). Effects of reserpine, segontin and phenoxybenzamine on the catecholamines and ATP of isolated nerve and adrenal medullary storage granules. *Life Sciences* **3**, 35–40.

Farmer, J. B. & Petch, B. (1963). Interaction of cocaine and tyramine on the isolated mammalian heart. *J. Pharm. Pharmac.* **15**, 639–643.

Farrant, J. (1963). Interactions between cocaine, tyramine and noradrenaline at the noradrenaline store. *Br. J. Pharmac. Chemother.* **20**, 540–549.

Farrant, J., Harvey, J. A. & Pennefather, J. N. (1964). The influence of phenoxybenzyamine on the storage of noradrenaline in rat and cat tissues. *Br. J. Pharmac. Chemother.* **22**, 104–112.

Fawaz, G. & Simaan, J. (1963). Cardiac noradrenaline stores. *Br. J. Pharmac. Chemother.* **20**, 569–578.

Fischer, J. E. & Snyder, S. H. (1966). Disposition of norepinephrine-H^3 in sympathetic ganglia. *J. Pharmac. exp. Ther.* **150**, 190–195.

Fleckenstein, A. & Bass, H. (1953). Die Sensibilisierung der Katzen-Nickhaut für Sympathomimetica der Brenzkatechen-Reihe. *Arch. exp. Path. Pharmak.* **220**, 143–156.

Fleckenstein, A. & Burn, J. H. (1953). The effect of denervation on the action of sympathomimetic amines on the nictitating membrane. *Br. J. Pharmac. Chemother.* **8**, 69–78.

Fleckenstein, A. & Stöckle, D. (1955). Die Hemmung der Neuro-sympathomimetica durch Cocain. *Arch. exp. Path. Pharmak.* **224**, 401–415.

Fleming, W. W. & Schmidt, J. L. (1962). The sensitivity of the isolated rabbit ileum to sympathomimetic amines following reserpine pretreatment. *J. Pharmac. exp. Ther.* **135**, 34–38.

Fleming, W. W. & Trendelenburg, U. (1961). Development of supersensitivity to norepinephrine after pretreatment with reserpine. *J. Pharmac. exp. Ther.* **133**, 41–51.

Fröhlich, A. & Loewi, O. (1910). Über eine Steigerung der Adrenalinempfindlichkeit durch Cocain. *Arch. exp. Path. Pharmak.* **62**, 159–169.

Furchgott, R. F. (1955). The pharmacology of vascular smooth muscle. *Pharmac. Rev.* **7**, 183–265.

Furchgott, R. F., Kirpekar, S. M., Rieker, M. & Schwab, A. (1963). Actions and interactions of norepinephrine, tyramine and cocaine on aortic strips of rabbit and left atria of guinea pig and cat. *J. Pharmac. exp. Ther.* **142**, 39–58.

Gaddum, J. H. & Kwiatkowski, H. (1938). The action of ephedrine. *J. Physiol., Lond.* **94**, 87–100.

Garattini, S. & Valzelli, L. (1958). Researches on the mechanism of reserpine sedative action. *Science* **128**, 1278–1279.

Gillespie, J. S. & Kirpekar, S. M. (1965). The inactivation of infused noradrenaline by the cat spleen. *J. Physiol., Lond.* **176**, 205–227.

Gillis, C. N. (1963). Increased retention of exogenous norepinephrine by cat atria after electrical stimulation of the cardioaccelerator nerves. *Biochem. Pharmac.* **12**, 593–595.

Gillis, C. N. (1965). Altered cardiac retention of exogenous noradrenaline produced by stress in young rabbits. *Nature, Lond.* **207**, 1302–1304.

Glowinski, J., Axelrod, J., Kopin, I. J. & Wurtman, R. J. (1964). Physiological disposition of H^3-norepinephrine in the developing rat. *J. Pharmac. exp. Ther.* **146**, 48–53.

Glowinski, J., Iversen, L. L. & Axelrod, J. (1966). Storage and synthesis of norepinephrine in the reserpine-treated rat brain. *J. Pharmac. exp. Ther.* **151**, 385–399.

Haefely, W., Hürlimann, A. & Thoenen, H. (1964). A quantitative study of the effect of cocaine on the response of the nictitating membrane to nerve stimulation and to injected noradrenaline. *Br. J. Pharmac. Chemother.* **22**, 5–21.

Hamberger, B., Malmfors, T., Norberg, K.-A. & Sachs, C. (1964). Uptake and accumulation of catecholamines in peripheral adrenergic neurones of reserpinized animals, studied with a histochemical method. *Biochem. Pharmac.* **13**, 841–844.

Hardman, J. G. & Mayer, S. E. (1965). The influence of cocaine on some metabolic effects and the distribution of catecholamines. *J. Pharmac. exp. Ther.* **148**, 29–39.

Hertting, G. (1965). The effect of drugs and sympathetic denervation on noradrenaline uptake and binding in animal tissues. In *Pharmacology of Cholinergic and Adrenergic Transmission* (ed. Douglas and Carlsson), pp. 277–288. Oxford: Pergamon Press.

Hertting, G., Axelrod, J. & Patrick, R. W. (1962). Actions of bretylium and guanethidine on the uptake and release of H^3-noradrenaline. *Br. J. Pharmac. Chemother.* **18**, 161–166.

Hertting, G., Axelrod, J. & Whitby, L. G. (1961). Effect of drugs on the uptake and metabolism of H^3-norepinephrine. *J. Pharmac. exp. Ther.* **134**, 146–153.

Hertting, G. & Schiefthaler, T. (1963). Beziehung zwischen Durchflussgrosse und Noradrenalin Freisetzung bei Nervenreizung der isoliert durchströmten Katzenmilz. *Arch. exp. Path. Pharmak.* **246**, 13–14.

Hillarp, N.-Å. & Malmfors, T. (1964). Reserpine and cocaine blocking of the uptake and storage mechanisms in adrenergic nerves. *Life Sciences* **3**, 703–708.

Holtz, P., Osswald, W. & Stock, K. (1960). Über die Beeinflussung der Wirkungen sympathicomimetischer Amine durch Cocain und Reserpin. *Arch. exp. Path. Pharmak.* **239**, 14–28.

Holzbauer, M. & Vogt, M. (1956). Depression by reserpine of the noradrenaline concentration in hypothalamus of the cat. *J. Neurochem.* **1**, 8–11.

Huković, S. (1959). Isolated rabbit atria with sympathetic supply. *Br. J. Pharmac. Chemother.* **14**, 372–376.

Huković, S. & Muscholl, E. (1962). Die Noradrenalin-Abgabe aus dem isolierten Kaninchenherzen bei sympathischer Nervenreizung und ihre pharmakologische Beeinflussung. *Arch. exp. Path. Pharmak.* **244**, 81–96.

Innes, I. R. & Kosterlitz, H. W. (1954). The effects of preganglionic and postganglionic denervation on the responses of the nictitating membrane to sympathomimetic substances. *J. Physiol., Lond.* **124**, 25–43.

Innes, I. R. & Krayer, O. (1958). Studies on veratrum alkaloids. XXVII. The negative chronotropic action of veratramine and reserpine in the heart depleted of catechol amines. *J. Pharmac. exp. Ther.* **124**, 245–251.

Isaac, L. & Goth, A. (1965). Interaction of antihistaminics with norepinephrine uptake. A cocaine-like effect. *Life Sciences* **4**, 1899–1904.

Iversen, L. L. (1963). The uptake of noradrenaline by the isolated perfused rat heart. *Br. J. Pharmac. Chemother.* **21**, 523–537.

Iversen, L. L. (1964). The inhibition of noradrenaline uptake by sympathomimetic amines. *J. Pharm. Pharmac.* **16**, 435–437.

Iversen, L. L. (1965*a*). The inhibition of noradrenaline uptake by drugs. *Adv. Drug Res.* **2**, 5–23.

Iversen, L. L. (1965*b*). The uptake of catecholamines at high perfusion concentrations in the isolated rat heart: a novel catecholamine uptake process. *Br. J. Pharmac.* **25**, 18–33.

Iversen, L. L. (1966). Accumulation of α-methyltyramine by the noradrenaline uptake process in the isolated rat heart. *J. Pharm. Pharmacol.* **18,** 481–484.

Iversen, L. L., Glowinski, J. & Axelrod, J. (1965). Uptake and storage of H^3-norepinephrine in reserpine pretreated rat heart. *J. Pharmac. exp. Ther.* **150**, 173–183.

Jonasson, J., Rosengren, E. & Waldeck, B. (1964). On the effects of some pharmacologically active amines on the uptake of arylalkylamines by adrenal medullary granules. *Acta physiol. scand.* **60**, 136–140.

Juorio, A. V. & Vogt, M. (1965). The effect of prenylamine on the metabolism of catechol amines and 5-hydroxytryptamine in brain and adrenal medulla. *Br. J. Pharmac. Chemother.* **24**, 566–573.

Kärki, N. T., Paasonen, M. K. & Vanhakartano, P. A. (1959). The influence of pentolonium, isoraunescine and yohimbine on the noradrenaline depleting action of reserpine. *Acta pharmac. tox.* **16**, 13–19.

Kirpekar, S. M. & Cervoni, P. (1963). Effect of cocaine, phenoxybenzamine and phentolamine on the catecholamine output from spleen and adrenal medulla. *J. Pharmac. exp. Ther.* **142**, 59–70.

Kirshner, N. (1962). Uptake of catecholamines by a particulate fraction of the adrenal medulla. *J. biol. Chem.* **237**, 2311–2317.

Kirshner, N., Rorie, M. & Kamin, D. (1963). Inhibition of dopamine uptake *in vitro* by reserpine administered *in vivo*. *J. Pharmac. exp. Ther.* **141**, 285–289.

Kopin, I. J. & Gordon, E. K. (1962). Metabolism of norepinephrine-H^3 released by tyramine and reserpine. *J. Pharmac. exp. Ther.* **138**, 351–359.

Kopin, I. J. & Gordon, E. K. (1963). Metabolism of administered and drug released norepinephrine-7-H^3 in the rat. *J. Pharmac. exp. Ther.* **140**, 207–216.

Kopin, I. J., Hertting, G. & Gordon, E. K. (1962). Fate of norepinephrine-H^3 in the isolated perfused rat heart. *J. Pharmac. exp. Ther.* **138**, 34–40.

Laverty, R. & Sharman, D. F. (1965). Modification by drugs of the metabolism of 3,4-dihydroxyphenylethylamine, noradrenaline and 5-hydroxytryptamine in the brain. *Br. J. Pharmac. Chemother*, **24**, 759–772.

Laverty, R., Sharman, D. F. & Vogt, M. (1965). Action of 2,4,5-trihydroxyphenylethylamine on the storage and release of noradrenaline. *Br. J. Pharmac. Chemother.* **24**, 549–560.

Leduc, J. (1961). Catecholamine production and release in exposure and acclimation to cold. *Acta physiol. scand.* **53**, Suppl. 183.

Lindmar, R. & Muscholl, E. (1961). Die Wirkung von Cocain, Guanethidin, Reserpin, Hexamethonium, Tetracain, und Psicain auf die Noradrenalin-Freisetzung aus dem Herzen. *Arch. exp. Path. Pharmak.* **242**, 214–227.

Lindmar, R. & Muscholl, E. (1962). Einfluss von Pharmaka auf die Elimination von Noradrenalin aus der Perfusionsflussigkeit des isolierten Herzen. *Arch exp. Path. Pharmak.* **243**, 347.

Lindmar, R. & Muscholl, E. (1964). Die Wirkung von Pharmaka auf die Elimination von Noradrenalin aus der Perfusionsflussigkeit und die Noradrenalin-Aufnahme in das isolierte Herz. *Arch. exp. Path. Pharmak.* **247**, 469–492.

Lindmar, R. & Muscholl, E. (1965). Die Verstärkung der Noradrenalinwirkung durch Tyramin. *Arch. exp. Path. Pharmak.* **252**, 122–133.

Lippmann, W. & Wishnick, M. (1965). Effects of the administration of epsilon amino caproic acid on catecholamine and serotonin levels in the rat and dog. *J. Pharmac. exp. Ther.* **150**, 196–202.

McLean, J. R. & Cohen, F. (1963). The release of epinephrine from rabbit adrenal medulla chromaffin granules. *Life Sciences* **2**, 261–265.

Mackenna, B. R. (1965). Uptake of catecholamines by the hearts of rabbits treated with segontin. *Acta physiol. scand.* **63**, 413–422.

Macmillan, W. H. (1959). A hypothesis concerning the effect of cocaine on the action of sympathomimetic amines. *Br. J. Pharmac. Chemother.* **14**, 385–391.

Malmfors, T. (1965). Studies on adrenergic nerves. *Acta physiol. scand.* **64**, Suppl. 248.

Maxwell, R. A., Plummer, A. J., Povalski, H., Schneider, F. & Coombs, H. (1959). A comparison of some of the cardiovascular actions of methylphenidate and cocaine. *J. Pharmac. exp. Ther.* **126**, 250–257.

Merritt, J. H., Schultz, E. J. & Wykes, A. A. (1965). Effect of decaborane on norepinephrine content of rat brain. *Biochem. Pharmac.* **13**, 1364–1365.

Mirkin, B. L., Giarman, N. J. & Friedman, D. (1963). Uptake of noradrenaline by subcellular particles in homogenates of rat brain. *Biochem. Pharmac.* **12**, 214–216.

Mueller, R. A. & Shideman, F. E. (1962). Direct evidence for release of myocardial norepinephrine by certain sympathomimetic amines. *Fedn. Proc.* **21**, 178.

Muskus, A. J. (1964). Evidence for different sites of action of reserpine and guanethidine. *Arch. exp. Path. Pharmak.* **248**, 498–513.

Musacchio, J. M., Kopin, I. J. & Weise, V. K. (1965). Subcellular distribution of some sympathomimetic amines and their β-hydroxylated derivatives in the rat heart. *J. Pharmac. exp. Ther.* **148**, 22–28.

Muscholl, E. (1960). Die Hemmung der Noradrenalin-Aufnahme des Herzens durch Reserpin und die Wirkung von Tyramin. *Arch. exp. Path. Pharmak.* **240**, 234–241.

Muscholl, E. (1961). Effect of cocaine and related drugs on the uptake of noradrenaline by heart and spleen. *Br. J. Pharmac. Chemother.* **16**, 352–359.

Muscholl, E. & Weber, E. (1965). Die Hemmung der Aufnahme von α-Methylnoradrenalin in das Herz durch sympathomimetische Amine. *Arch. exp. Path. Pharmak.* **252**, 134–143.

Nickerson, M. (1965). Drugs inhibiting adrenergic nerves and structures innervated by them. In *The Pharmacological Basis of Therapeutics*, 3rd ed. (ed. Goodman and Gillman), pp. 546–577. New York: Macmillan.

Nikkilä, E. A., Torsti, P. & Penttilä, O. (1965). Effects of fasting, excercise and reserpine on catecholamine content and lipoprotein lipase activity of rat heart and adipose tissue. *Life Sciences* **4**, 27–35.

Oliverio, A. & Stjärne, L. (1965). Acceleration of noradrenaline turnover in the mouse heart by cold exposure. *Life Sciences* **4**, 2339–2343.

Pletscher, A. & Gey, K. F. (1961). Influence of chlorpromazine and chlorprothixene on the cerebral metabolism of 5-hydroxytryptamine, norepinephrine and dopamine. *J. Pharmac. exp. Ther.* **133**, 18–24.

Pletscher, A., Brossi, A. & Gey, K. F. (1962). Benzoquinolizine derivatives, a new class of monoamine decreasing drugs with psychotropic action. *Int. Rev. Neurobiol.* **6**, 275–306.

Porter, C. C., Totaro, J. A. & Stone, C. A. (1963). The effect of 6-hydroxydopamine and some other compounds on the concentration of norepinephrine in the hearts of mice. *J. Pharmac. exp. Ther.* **140**, 308–316.

Porter, C. C., Totaro, J. A. & Burcin, A. (1965). The relationship between radioactivity and norepinephrine concentrations in the brains and hearts of mice following administration of labelled α-methyldopa or 6-hydroxydopamine. *J. Pharmac. exp. Ther.* **150**, 17–22.

Potter, L. T. & Axelrod, J. (1963). Properties of norepinephrine storage particles of the rat heart. *J. Pharmac. exp. Ther.* **142**, 299–305.

REFERENCES

Quinn, G. P., Shore, P. A. & Brodie, B. B. (1959). Biochemical and pharmacological studies of Ro 1–9569 (Tetrabenazine), a non-indole tranquillizing agent with reserpine-like effects. *J. Pharmac. exp. Ther.* **127**, 103–109.

Rosell, S. & Sedvall, G. (1961). Restoration of vasoconstriction effects in reserpinized cats. *Acta physiol. scand.* **53**, 174–184.

Rosell, S., Kopin, I. J. & Axelrod, J. (1963). Fate of H^3-norepinephrine in skeletal muscle before and following sympathetic stimulation. *Am. J. Physiol.* **205**, 317–321.

Ross, S. B. & Renyi, A. L. (1965). Blocking action of sympathomimetic amines on the uptake of tritiated noradrenaline by mouse cerebral cortex tissues *in vitro*. *Acta pharmac. tox.* **21**, 226–239.

Schapiro, S. (1958). Effect of a catecholamine blocking agent (dibenzyline) on organ content and urine excretion of noradrenaline and adrenaline. *Acta physiol. scand.* **42**, 371–375.

Schmidt, J. L. & Fleming, W. W. (1963). The structure of sympathomimetics as related to reserpine induced sensitivity changes in the rabbit ileum. *J. Pharmac. exp. Ther.* **139**, 230–237.

Schöne, H. H. & Lindner, E. (1960). Die Wirkungen des *N*-[3′-phenyl-propyl-(2′)]-1,1-diphenylpropyl-(3)-amins auf den Stoffwechsel von Serotonin und Noradrenalin. *Arzneimittel-Forsch.* **10**, 583–585.

Schümann, H. J. & Philippu, A. (1962). Release of catechol amines from isolated medullary granules by sympathomimetic amines. *Nature, Lond.* **193**, 890–891.

Schümann, H. J. & Wegmann, E. (1960). Über den Angriffspunkt der inderekten Wirkung Sympathicomimetischer. *Arch. exp. Path. Pharmak.* **240**, 275–284.

Shore, P. A. (1962). Release of serotonin and catecholamines by drugs. *Pharmac. Rev.* **14**, 531–550.

Sigg, E. B., Soffer, L. & Gyermek, L. (1963). Influence of imipramine and related psychoactive agents on the effect of 5-hydroxytryptamine and catecholamines on the cat nictitating membrane. *J. Pharmac. exp. Ther.* **142**, 13–20.

Stafford, A. (1963). Potentiation of some catechol amines by phenoxybenzamine, guanethidine and cocaine. *Br. J. Pharmac. Chemother.* **21**, 361–367.

Stjärne, L. (1961). Tyramine effects on catechol amine release from spleen and adrenals in the cat. *Acta physiol. scand.* **51**, 224–229.

Stjärne, L. (1964). Studies of catecholamine uptake, storage and release mechanisms. *Acta physiol. scand.* **62**, Suppl. 228.

Stone, C. A., Porter, C. C., Stavorski, J. M., Ludden, C. T. & Totaro, J. A. (1964). Antagonism of certain effects of catecholamine-depleting agents by antidepressant and related drugs. *J. Pharmac. exp. Ther.* **144**, 196–204.

Sulser, F. & Brodie, B. B. (1960). Is reserpine tranquillization linked to change in brain serotonin or brain norepinephrine? *Science* **131**, 1440–1441.

Swaine, C. R. (1963). Effect of adrenergic blocking agents on tyramine-induced release of catecholamines from the cat heart. *Proc. Soc. exp. Biol. Med.* **112**, 388–390.

Tainter, M. L. (1929). Comparative effects of ephedrine and epinephrine on blood pressure, pulse and respiration with reference to their alteration by cocaine. *J. Pharmac. exp. Ther.* **36**, 569–594.

Tainter, M. L. & Chang, D. K. (1927). The antagonism of the pressor action of tyramine by cocaine. *J. Pharmac. exp. Ther.* **30**, 193–207.

Thoenen, H., Hürlimann, A. & Haefely, W. (1964*a*). The effect of sympathetic nerve stimulation on volume, vascular resistance and norepinephrine output in the isolated perfused spleen of the cat and its modification by cocaine. *J. Pharmac. exp. Ther.* **143**, 57–63.

Thoenen, H., Hürlimann, A. & Haefely W. (1964*b*). Wirkungen von Phenoxybenzamin, Phentolamin und Azapetin auf adrenergische Synapsen der Katzenmilz. *Helv. Physiol. Pharmac. Acta* **22**, 148–161.

Thoenen, H., Hürlimann, A. & Haefely, W. (1964*c*). Mode of action of imipramine and 5-(3′-methylaminopropyliden)-dibenzo[a,e]cyclohepta[1,3,5]trien hydrochloride (RO 4–6011), a new antidepressant drug, on peripheral adrenergic mechanisms. *J. Pharmac. exp. Ther.* **144**, 405–414.

Titus, E. O. & Spiegel, H. E. (1962). Effect of desmethylimipramine (DMI) on uptake of norepinephrine-7-H^3 (NE) in heart. *Fedn. Proc.* **21**, 179.

Trendelenburg, U. (1959). The supersensitivity caused by cocaine. *J. Pharmac. exp. Ther.* **125**, 55–65.

Trendelenburg, U. (1962). Restoration by sympathomimetics of the response of isolated atria of reserpine-treated guinea-pigs to tyramine and DMPP. *Fedn. Proc.* **21**, 322.

Trendelenburg, U. (1963). Supersensitivity and subsensitivity to sympathomimetic amines. *Pharmac. Rev.* **15**, 225–276.

Trendelenburg, U. (1966). Mechanisms of supersensitivity and subsensitivity to sympathomimetic amines. (Second International Catecholamine Symposium.) *Pharmac. Rev.* **18**, 629–640.

Trendelenburg, U. & Pfeffer, R. I. (1964). Effects of infusions of sympathomimetic amines on the response of spinal cats to tyramine and to sympathetic stimulation. *Arch. exp. Path. Pharmak.* **248**, 39–53.

Trendelenburg, U., Muskus, A., Fleming, W. W. & Gomez, B. (1962*a*). Modification by reserpine of the action of sympathomimetic amines in spinal cats; a classification of sympathomimetic amines. *J. Pharmac. exp. Ther.* **138**, 170–180.

Trendelenburg, U., Muskus, A., Fleming, W. W. & Gomez, B. (1962*b*). Effect of cocaine, denervation and decentralization on the response of the nictitating membrane to various sympathomimetic amines. *J. Pharmac. exp. Ther.* **138**, 181–193.

Weil-Malherbe, H. & Posner, H. S. (1963). The effect of drugs on the release of epinephrine from adrenomedullary particles *in vitro*. *J. Pharmac. exp. Ther.* **140**, 93–102.

Weiner, N., Draskoćzy, P. R. & Burack, M. R. (1962). The ability of tyramine to liberate catecholamines *in vivo*. *J. Pharmac. exp. Ther.* **137**, 47–55.

Weiner, N., Perkins, M. & Sidman, R. L. (1962). Effect of reserpine on noradrenaline content of innervated and denervated brown adipose tissue of the rat. *Nature, Lond.* **193**, 137–138.

Whitby, L. G., Hertting, G. & Axelrod, J. (1960). Effect of cocaine on the disposition of noradrenaline labelled with tritium. *Nature, Lond.* **187**, 604–605.

Wurtman, R. J., Kopin, I. J. & Axelrod, J. (1963). The disposition of catecholamines in the rat uterus and the effect of drugs and hormones. *J. Pharmac. exp. Ther.* **144**, 150–155.

Wurtman, R. J., Kopin, I. J. & Axelrod, J. (1964). Thyroid function and the cardiac disposition of catecholamines. *Endocrinology* **73**, 63–74.

Zaimis, E. (1964). Pharmacology of the autonomic nervous system. *Ann. Rev. Pharmac.* **4**, 365–400.

9

NORADRENALINE STORAGE POOLS—FACT OR ARTIFACT?

In recent years there has been an increasing awareness that the noradrenaline store in postganglionic sympathetic nerve endings cannot be regarded as a single homogeneously mixed pool. While attempts to view the sympathetic nerve ending as a highly complex and varied entity are wholly laudable, the current proliferation of noradrenaline storage pools has probably gone too far. Instead of simplifying our concepts about the adrenergic transmitter stores, the introduction of a multiplicity of hypothetical storage pools has tended to confuse the problem still further. Much of the confusion has arisen because the term 'pool' may have several different connotations. In one sense, and perhaps most correctly, pools are entirely abstract concepts—defined purely on kinetic or other mathematical grounds. This abstract definition, however, is often confused with, or merges with, the view that a 'pool' implies some morphologically distinct, physically delineated compartment. Generally one pool of molecules is distinguished from another pool by the existence of circumstances which prevent free mixing between the molecules in the two pools. In biological systems, such diffusion barriers generally imply the existence of a lipoprotein membrane of limited permeability to the substance in question. However, it is also possible to envisage pools of molecules within the same cell which are not separated by any physical barrier. Thus, in an adrenergic neurone, molecules at two points along the axon which are physically separated by a distance of some centimetres will not be able to mix with one another freely, although they may be present in free solution in the axoplasm.

1. Anatomical and morphological considerations

In the sense that noradrenaline is found in all parts of the postganglionic sympathetic neurone, there are clearly several morphologically distinct compartments of transmitter: in the perikaryon, the preterminal axon and the terminal fibres of the adrenergic ground plexus. Since it is probably only the latter stores which are released in response to nerve stimulation, the terminal stores can be differentiated in terms of function as well as morphology from the other two. There is evidence that the transmitter is distributed very unevenly between these three compartments. Norberg & Hamberger (1964) have estimated the concentrations of noradrenaline to be 10–100 μg/g in sympathetic ganglion cells, 100–500 μg/g in preterminal axons and approximately 10000 μg/g in sympathetic terminals, on the basis of fluorescence microscopy.

There is also anatomical evidence suggesting that the transmitter is unevenly distributed within the terminal regions of the sympathetic neurones. As discussed in chapter 4, histochemical studies have revealed that the transmitter is highly concentrated in the varicosities which occur along the length of the terminal fibres, and is present in only small amounts in the connecting lengths of axon terminal between such varicosities. This concentration of noradrenaline at specific points along the length of the terminal fibre is correlated with the presence of numerous storage vesicles at these points. This raises the further question of whether all the noradrenaline in the terminal regions of the sympathetic neurone is bound in such particles, or whether the transmitter is distributed between two morphologically distinct pools: particle-bound and axoplasm. The histochemical technique for catecholamines lacks sufficient resolution to answer this question. The faint fluorescence observed in histochemical studies in sympathetic ganglion cells, preterminal axons and between varicosities in terminal fibres could well be due to the presence of noradrenaline in a sparsely distributed population of particles in such regions. The biochemical evidence on this point is also unconvincing. In all studies of the subcellular distribution of noradrenaline in sympathetic nerves a considerable proportion of the total noradrenaline content is recovered in the soluble supernatant on centrifugation of tissue homogenates. This does not prove the existence of a pool of 'free' noradrenaline in the axoplasm, however, since it has been argued that the supernatant catecholamine may arise from the destruction of some storage particles during the harsh procedures used to disrupt the tissue. On the other hand, it is somewhat surprising that several different laboratories have arrived at consistent and closely similar estimates of

Table 9.1. *A comparison of published values for the subcellular distribution of exogenous and endogenous noradrenaline in the hearts of various species*

Species and tissue	Endogenous or exogenous NA	Noradrenaline in microsomal fraction (% microsomal + supernatant)	Reference
Rat, heart	Endogenous and H^3-NA	*c.* 70	Potter & Axelrod, 1963
Rat, heart	Endogenous	56	Author's unpublished results
Dog, heart	Endogenous	74	Wegmann & Kako, 1961
Cat, left ventricle	Endogenous	54	Campos *et al.* 1963
Dog, right auricle	Endogenous	64	Campos & Shideman, 1962
Rabbit, heart	Endogenous	52	Euler & Lishajko, 1965
Guinea-pig, heart	Endogenous	66	Schümann *et al.* 1964
Rat, heart	H^3-NA	54	Iversen *et al.* 1965
Mouse, heart	H^3-NA	46	Iversen & Whitby, 1963
Rabbit, heart	H^3-NA	25	Stjärne, 1964
Rat, heart	H^3-NA	48	Musacchio *et al.* 1965

Endogenous and/or exogenous noradrenaline in microsomal storage particles as a percentage of the total amount of catecholamine remaining after removing unbroken tissue fragments by a preliminary low-speed centrifugation. Despite differences in experimental technique and species, the results obtained are remarkably consistent.

the ratio of particle-bound to free noradrenaline in adrenergic nerves (Table 9.1). If the release of particle-bound catecholamine into the homogenizing medium were entirely responsible for the existence of supernatant noradrenaline, it might be expected that this would be an artifact of widely different quantitative importance under different experimental conditions and would thus tend to give highly variable results in such studies. This, however, is not the case. The particle/supernatant ratio is also characteristically different for various structural analogues of noradrenaline. Musacchio, Kopin & Weise (1965) studied the subcellular distribution of several radioactively labelled analogues of noradrenaline in the sympathetic innervation of the rat heart (Table 9.2) and found that the particle/supernatant ratio

Table 9.2. *Subcellular distribution of various sympathomimetic amines in the rat heart and their release by tyramine*

(From Musacchio *et al.* 1965.)

Amine	Percentage recovered in microsomal particle fraction	Percentage released by tyramine*
Noradrenaline	48 ± 1·1	50
Dopamine†	38 ± 2·2	87
m-Octopamine	30 ± 3·4	91
α-Methyl octopamine	27 ± 3·1	—
Octopamine	25 ± 1·4	96
α-Methyl tyramine†	6 ± 0·5	—
Tyramine* and *m*-tyramine†	—‡	—

* For tyramine-release experiments animals were killed 2 h after the administration of the labelled amines and tyramine (10 mg/kg, i.p.) was administered 1 h after the amine administration; values are the means for six experiments.

† Subcellular distribution of non-β-hydroxylated amines was determined in animals in which β-hydroxylation was inhibited by treatment with disulfiram.

‡ Amounts of amine in particulate fraction too low to be accurately measured.

Tritium-labelled amines or their precursors were administered intravenously; 1 h later rat hearts were homogenized in isotonic sucrose, centrifuged at 20000*g* for 10 min (pellet of unbroken cells and debris discarded); the distribution of radioactive amine between soluble supernatant and microsomal pellet was then determined after centrifugation at 100000*g* for 60 min. Results are the means for 3–6 determinations ± S.E.M.

varied considerably among the different substances, being highest for noradrenaline, lower for dopamine and octopamine and very low for non-β-hydroxylated amines such as tyramine.

These results suggest that noradrenaline is normally distributed between two distinct pools: in the axoplasm, in free solution; and in a bound form within storage particles. However, it might also be argued that the supernatant fraction of noradrenaline merely represents a portion of the amine normally bound in particles which is especially susceptible to release. This would imply that in the intact neurone all the noradrenaline is present within storage particles, but that these particles contain two forms of stored amine, one firmly bound and the other readily releasable. Unfortunately there seems to be no obvious experimental approach to resolve this question. The uncertainty which is inherent in studies of the subcellular distribution of catecholamines is no new phenomenon to biochemists. For example, the question of whether cytochrome *c* is localized exclusively within mitochondria cannot be resolved by centrifugation techniques, since this compound rapidly leaks out of mitochondria when they are exposed to the homogenizing medium.

2. Distribution of exogenous noradrenaline in the transmitter store

There is considerable evidence that exogenous noradrenaline is not taken up into a single homogeneous store of noradrenaline in sympathetic nerves (see chapter 7). Only a limited exchange of exogenous noradrenaline with tritiated noradrenaline occurs when tissues are exposed to the labelled amine. In studies in the isolated heart of the rat, not more than 25–30 % of the endogenous noradrenaline store exchanged with labelled noradrenaline during continuous 30–60 min perfusions with the labelled amine (Iversen, 1963). Similarly limited exchange of the endogenous transmitter store with exogenous noradrenaline has been reported by Dengler, Spiegel & Titus (1961) and by Crout (1964). In isolated adrenergic nerve granules, however, as much as 75–80 % of the endogenous noradrenaline content exchanges quite rapidly with exogenous noradrenaline when the particles are incubated in labelled noradrenaline (Euler, Stjärne & Lishajko, 1963; Stjärne, 1964). It would thus appear that in the nerve storage granules and particularly in the intact tissue the endogenous noradrenaline content is stored in such a way as to prevent free mixing with exogenous catecholamines which are taken up into the cells or particles.

In the isolated rat heart the curve of the rate of noradrenaline uptake against time during a continuous perfusion with labelled noradrenaline could be resolved into two components (Fig. 7.8) (Iversen, 1963). The accumulation of noradrenaline thus appears to occur in at least two compartments within the sympathetic nerve. Compartment I fills very rapidly, and appears to contain only negligible amounts of endogenous noradrenaline, since little or no exchange with endogenous noradrenaline occurs at this time. A second compartment, II, fills more slowly, and contains at least 25–30 % of the endogenous noradrenaline content, since exchange of labelled amine with endogenous noradrenaline occurs during the filling of this compartment. The remaining 75 % of the endogenous catecholamine content may be in compartment II or in a separate compartment (III). Compartment I may correspond to a small pool of exogenous noradrenaline which accumulates in the axoplasm during the uptake of exogenous catecholamine. The uptake of noradrenaline occurs only into compartment I in reserpine-pretreated animals, in which particle binding sites are destroyed. In this sense compartment I is an artifact, since it contains only small amounts of endogenous noradrenaline, or exogenous noradrenaline immediately after it enters the neurone. Compartment II may represent noradrenaline in storage particles; there is no accumulation of exogenous noradrenaline in this compartment in reserpinized tissues.

An uneven distribution of exogenous noradrenaline within the sympathetic neurones has also been inferred from the complex nature of the disappearance of the accumulated catecholamine from tissues (see chapter 7). Two or three phases can usually be detected in the disappearance of accumulated H^3-noradrenaline from tissues. An initial very rapid phase (half time *c.* 15 min) is seen only when high doses of noradrenaline are administered (Crout, 1964). A second phase with a half time of *c.* 5–6 h is commonly observed, followed by a very slow phase of disappearance with half time of *c.* 24 h (Fig. 7.7). However, if tracer doses of labelled noradrenaline are administered, the disappearance of the accumulated amine may be monophasic (Costa *et al.* 1966). The situation is further complicated by the fact that most studies have made use of a racemic mixture of DL-H^3-noradrenaline. While there is little doubt that the L-enantiomer is taken up more rapidly than the D-enantiomer, there is some doubt as to the subsequent fate of the two enantiomers once they have entered sympathetic nerves. Kopin & Bridgers (1963) and Maickel, Beaven & Brodie (1963) have reported that D-noradrenaline disappears from tissues more rapidly than L-noradrenaline, but the results of Andén (1964) and Crout (1964) do not seem to substantiate this view. If it is true that D-noradrenaline disappears more rapidly from tissues than L-noradrenaline, then the rapid disappearance of the D-enantiomer after an injection of DL-H^3-noradrenaline might account for one of the two initial rapid phases of disappearance described above. The existence of a very rapid phase of disappearance after the administration of large doses of DL-H^3-noradrenaline could perhaps be explained on this basis—since the uptake of the D-enantiomer is quantitatively more important as the administered dose of DL-noradrenaline is increased (see chapter 7). It is unlikely that the second phase of disappearance can be explained on this basis, since this phase may be observed even after the administration of small doses of L-H^3-noradrenaline (Costa *et al.* 1966).

In summary, there is considerable evidence to suggest that exogenous catecholamines do not mix freely with the endogenous noradrenaline content when they are taken up into sympathetic nerves, and this in turn suggests the existence of compartments within the transmitter store.

3. Depletion of noradrenaline stores by drugs

The noradrenaline content of sympathetic nerves can be severely reduced by drug treatment without a comparable reduction in the response of the tissue to nerve stimulation or to indirectly acting sympathomimetic amines such as tyramine. The experiments of Crout, Muskus & Trendelenburg (1962) and Muskus (1964) are elegant examples of this type of study. Crout *et al.* (1962) observed that after the administration of increasing amounts of reserpine the noradrenaline content of guinea-pig atria was progressively reduced. The response of the tissue to tyramine, however, was little affected by doses of reserpine which depleted 50 % of the noradrenaline content. The tyramine response was only markedly impaired when the noradrenaline content was reduced to less than 10 % of the normal level (Fig. 9.1). On the basis of this evidence the authors suggested that two storage compartments of noradrenaline exist in sympathetic nerves. One store of 'bound' noradrenaline, which contains most of the noradrenaline and can be depleted by reserpine, is not necessary for a normal tyramine response. A second 'available' pool of noradrenaline is the one directly involved in the tyramine response; this store may not be depleted by reserpine, but eventually empties in the presence of reserpine when the reservoir of 'bound' noradrenaline (from which the 'available' pool is replenished) has been completely depleted. Similarly Andén & Magnusson (1965) and Carlsson (1964) have shown that the noradrenaline content of tissues can be considerably reduced by the introduction of α-methyl noradrenaline, metaraminol or D-adrenaline into the stores, without appreciably diminishing the effects of sympathetic

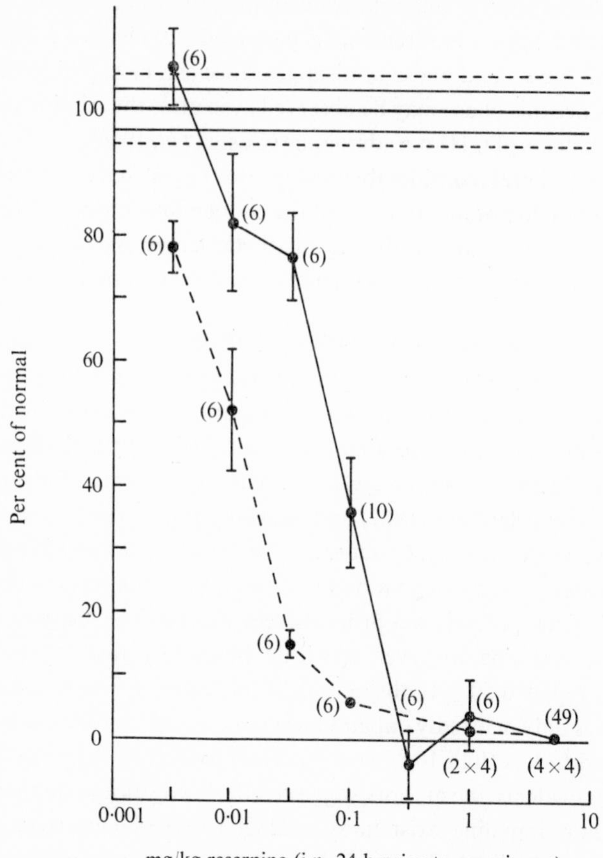

Fig. 9.1. Effect of pretreatment with graded doses of reserpine (given intraperitoneally 24 h prior to experiment) on the maximal response of isolated guinea-pig atria to tyramine and on their noradrenaline content. Solid line: response of rate of beat to 10^{-5} tyramine expressed as percentage of control observations. Broken line: noradrenaline content in percentage of control observations. The horizontal lines near the 100 % level indicate the standard error of the mean of the control observations (16 for response to tyramine, 8 for noradrenaline content); standard errors of means are indicated by vertical bars, numbers of observations in parentheses. (From Crout, Muskus & Trendelenburg, 1962.)

nerve stimulation. Since the α-methylated amines and D-adrenaline are less potent than noradrenaline at adrenergic receptors, the authors suggest that this effect cannot be explained simply by the hypothesis that the amines act as false transmitters, but implies that a small compartment of noradrenaline (which is all that is necessary for normal adrenergic function) remains intact after such drug treatment.

The experiments of Potter, Axelrod & Kopin (1962) and Potter & Axelrod (1963) demonstrated the distribution of H^3-noradrenaline in two storage pools in sympathetic nerves. One pool was readily released by tyramine, while the other appeared resistant to tyramine release. If tyramine was administered a short time after the injection of H^3-noradrenaline the drug released as much as 75 % of the accumulated catecholamine from the rat heart, while releasing only some 40 % of the endogenous catecholamine content. If tyramine was administered 24 or 48 h after the injection of H^3-noradrenaline, however, only 17 % of the labelled catecholamine was released from the heart. The specific activity of the H^3-noradrenaline released by tyramine was thus initially higher than that in the tissue as a whole, and subsequently lower (Fig. 9.2). Chidsey & Harrison (1963) reported somewhat similar results in the dog heart; in their experiments, however, the specific activity of H^3-noradrenaline released by tyramine at later times was equal to that in the tissue as a whole. These findings suggest that the labelled amine is present in at least two pools in the heart; one pool has a rapid turnover rate and is preferentially released by tyramine, and the other has a slower rate of turnover and is resistant to tyramine release. Unfortunately these results have often been misinterpreted to suggest the existence of two clearly distinguishable pools of noradrenaline in sympathetic nerves: one which *can* be released by tyramine and one which *cannot* be released by the drug. This is not the case, as shown recently by Neff, Tozer, Hammer & Brodie (1965) (Fig. 9.3). The whole of the noradrenaline content of the sympathetic nerves of the rat or guinea-pig heart can be depleted by the repeated administration of large doses of tyramine. These experiments, nevertheless, do not invalidate the findings of Potter *et al.* which still imply an uneven distribution of exogenous noradrenaline in storage pools in the rat heart. Although both postulated pools can be completely depleted by tyramine, one is preferentially released by the drug while the other is more resistant to

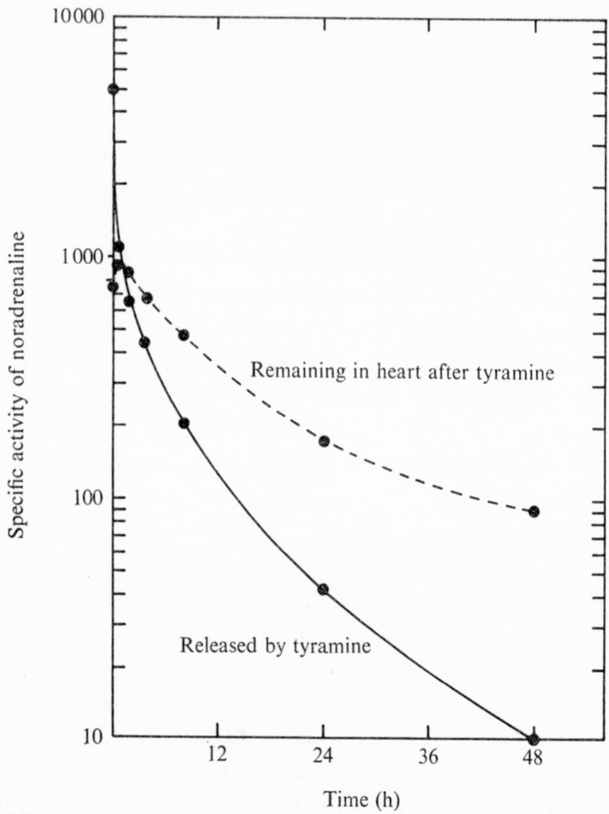

Fig. 9.2. Specific activity of H^3-noradrenaline released by tyramine from the rat heart. Rats were given H^3-noradrenaline (15 μc, i.v.) followed at various intervals by tyramine (10 mg/kg, i.m.). Fifteen minutes after the administration of tyramine hearts were analysed for noradrenaline and for H^3-noradrenaline. In all cases tyramine reduced the endogenous noradrenaline content from 0·83 to 0·61 μg/g, but at early times after H^3-noradrenaline tyramine released more of the labelled amine than at later times. The specific activity of tyramine-released noradrenaline was calculated from these data. Specific activities are expressed as mμc/μg noradrenaline; values are means for six animals. Zero time is taken from the point at which H^3-noradrenaline was injected. (From Potter & Axelrod, 1963.)

release. The two forms of stored noradrenaline may correspond to differences in the binding of the amine to storage particles. Musacchio *et al.* (1965) found that a single dose of tyramine could almost completely deplete amines such as octopamine or dopamine from the heart, while noradrenaline which is bound by both catechol and β-hydroxyl groups remained more resistant to tyramine release.

The development of tachyphylaxis to tyramine and other sympathomimetic amines has been related to the depletion of noradrenaline stores produced by the drug (Axelrod *et al.* 1962); this also does not mean that when tachyphylaxis has developed tyramine can no longer release noradrenaline. More probably tachyphylaxis develops when

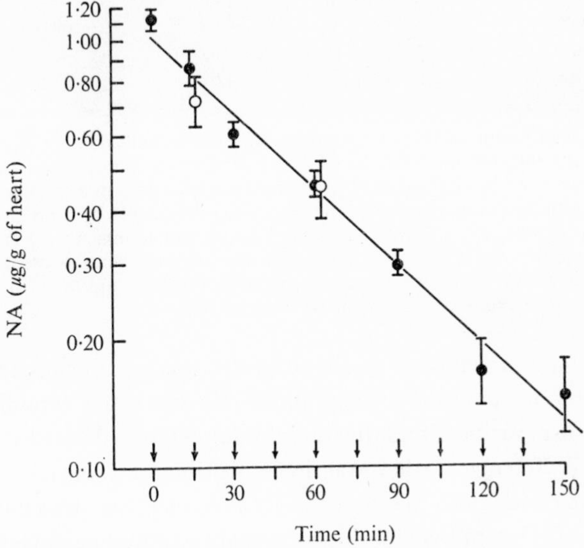

Fig. 9.3. The effects of tyramine on noradrenaline stores of rat heart during the infusion or the repeated intraperitoneal injection of the drug. Dosage schedule: ●, 50 mg/kg, i.p., at 15 min intervals for 150 min, each point is an average of five animals; ○, infusion of 150 mg/kg per 15 min for 60 min, each point is an average of three animals. Vertical bars indicate the standard error. The line was drawn according to the method of least squares. (From Neff *et al.* 1965.)

tyramine is no longer able to release sufficient noradrenaline to produce a pharmacological response. These results are in agreement with those of Trendelenburg (1961), Crout *et al.* (1962) and Trendelenburg (1963) who proposed the existence of a small pool of available noradrenaline which is preferentially released by tyramine and which can be replenished from a much larger pool, which is not immediately available for tyramine release.

Table 9.3. *Subcellular distribution of* H^3*-noradrenaline in normal and tyramine-depleted mouse hearts*

(From Iversen & Whitby, 1963.)

	Time after administration of H^3-noradrenaline	Percentage of H^3-NA in microsomal fraction	Percentage H^3-NA released by tyramine
Control	30 min	46·2	—
Tyramine	30 min	40·2	69·0
Control	24 h	45·8	—
Tyramine	24 h	42·1	23·5

Subcellular distribution of H^3-noradrenaline (NA) was examined either 30 min or 24 h after the intravenous injection of 0·2 μg H^3-noradrenaline. Results are expressed as the mean percentage of H^3-noradrenaline recovered in a fraction of microsomal particles—after removing unbroken tissue fragments by a preliminary low-speed centrifugation. Tyramine (15 mg/kg, i.p.) was administered 15 min before the animals were killed, the percentage of the total heart content of H^3-noradrenaline released by this treatment at the two times is indicated in the last column. Results are mean values for six experiments.

One possibility which these results suggested was that the distribution of H^3-noradrenaline between 'tyramine-releasable' and 'tyramine resistant' pools at various times after the injection of labelled noradrenaline might reflect changes in the intracellular distribution of the labelled catecholamine with time. We accordingly examined the subcellular distribution of H^3-noradrenaline in the normal and tyramine-depleted mouse heart at various times after the injection of H^3-noradrenaline. The results (Table 9.3) (Iversen & Whitby, 1963), however, failed to support this view. The distribution of H^3-noradrenaline between particulate and supernatant fractions in the mouse heart homogenates remained constant between 5 min and 24 h after the injection of H^3-noradrenaline. This finding has been confirmed by other workers (Stjärne, 1964; I. A. Michaelson, personal communication). Nevertheless, in confirmation of Potter *et al.* (1962), we found that at short times after injection a large proportion of the labelled amine could be released by tyramine, whereas a much smaller proportion of the accumulated amine could be released in this way 24 h after the injection of H^3-noradrenaline (Table 9.3). The subcellular distribution of H^3-noradrenaline was not significantly different in normal and tyramine-depleted hearts at the two times, indicating that tyramine did not preferentially release H^3-noradrenaline from either fraction. Similar results have been reported by Campos & Shideman (1962) and Campos, Stitzel & Shideman (1963). The uncertainty concerning the origin and size of the supernatant fraction in such experiments makes their interpretation difficult. However, it seems that 'tyramine-releasable' and 'tyramine-resistant' stores of noradrenaline do not correspond to any obvious morphological differences in the intraneuronal distribution of noradrenaline in the transmitter stores. Other authors have suggested that the two forms of noradrenaline in respect to tyramine release may correspond to noradrenaline in storage particles located in different topological relationships to the axonal membrane. According to this hypothesis, storage particles near the axonal membrane constitute the 'tyramine-releasable' pool; they accumulate exogenous H^3-noradrenaline preferentially because of their proximity to the membrane and are also preferentially attacked by tyramine for the same reason (Potter & Axelrod, 1963; Crout, 1964).

4. The refilling of noradrenaline stores in reserpine-depleted tissues

After the administration of reserpine the noradrenaline stores of sympathetic nerves are severely depleted and normal levels of noradrenaline are restored only after a long period (7–14 days). The ability of the tissues to respond to sympathetic nerve stimulation and to tyramine, however, recovers much more rapidly and may be almost normal 48 h after the administration of the drug, at a time when the noradrenaline stores are still severely depleted (Figs. 9.4, 9.5) (Andén, Magnusson & Waldeck, 1964; Muskus, 1964; Stjärne, 1964). This finding again suggests that only a small fraction of the total noradrenaline store is necessary for adrenergic transmission or for tyramine responses. Andén *et al.* (1964) were able to show that the recovery of adrenergic transmission in reserpine-treated tissues occurred at a time when the tissue recovered its ability to accumulate small amounts of exogenous H^3-noradrenaline. In the rat heart we were able to confirm this finding and to show that the recovery of the ability of the tissue to accumulate H^3-noradrenaline occurred at a time when the nerve storage particles regained the ability to store exogenous H^3-nor-

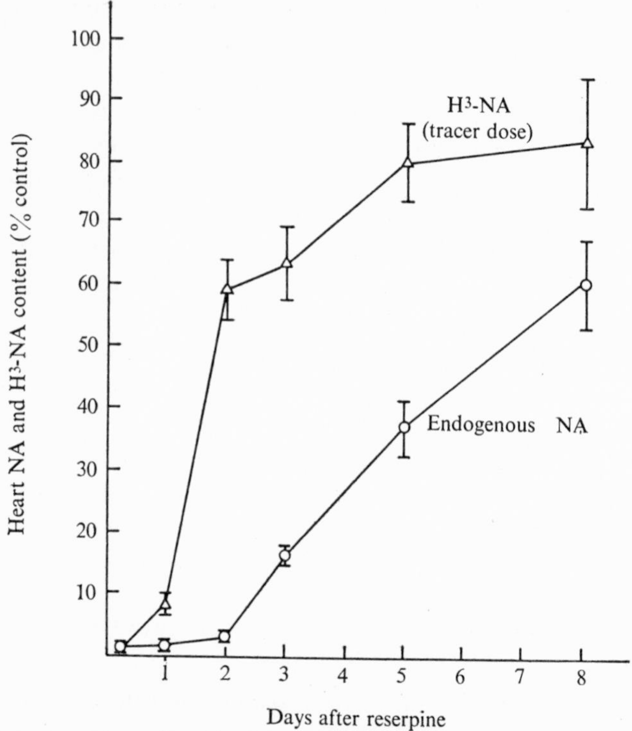

Fig. 9.4. Recovery of the ability to accumulate small amounts of H^3-noradrenaline and recovery of endogenous noradrenaline content of rat heart after the administration of a single dose of reserpine (5 mg/kg, i.p.). H^3-noradrenaline was administered (0·9 μg, i.v.) 20 min before the heart was taken for analysis of noradrenaline and H^3-noradrenaline content. Values are means $\pm$ standard errors of the mean and are expressed in percentages of control values obtained in normal animals. (From Iversen, Glowinski & Axelrod, 1965.)

adrenaline. At this time, however, the particles were only able to retain much smaller amounts of H^3-noradrenaline than normally, suggesting that the recovery of a small proportion of the normal population of storage particles was sufficient for a restoration of normal function (Iversen, Glowinski & Axelrod, 1965).

Many experiments have demonstrated that adrenergic transmission and the ability to respond to tyramine can be restored temporarily to reserpine-treated tissues by the exposure of the tissue to noradrenaline. This was first shown by Burn & Rand (1958), who later demonstrated that similar effects could be obtained by the administration of dopamine, L-dopa, *m*-tyrosine or phenylalanine (Burn & Rand, 1960). The 'refilling' of transmitter stores in reserpinized tissues has been the subject of many investigations (Muscholl, 1960; Kuschinsky, Lindmar, Lüllmann & Muscholl, 1960; Mirkin & Euler, 1963; Gillespie & Mackenna, 1961; Trendelenburg & Pfeffer, 1964; Trendelenburg & Crout, 1964). The restoration of tyramine actions and the response to sympathetic nerve stimulation in reserpine-treated tissues on exposure

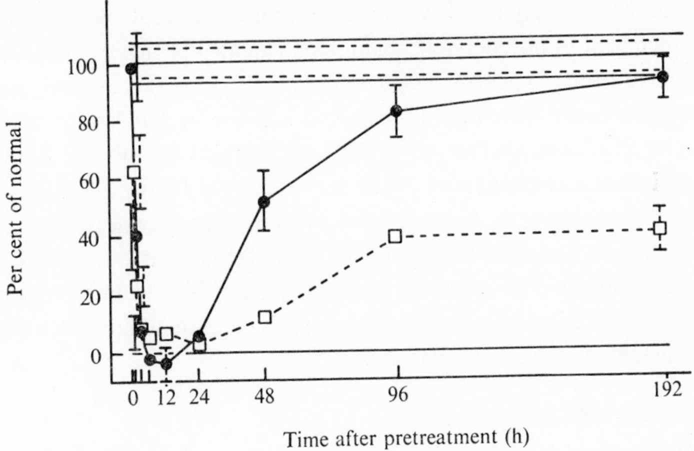

Fig. 9.5. Recovery of responsiveness to tyramine (—●—) and recovery of endogenous noradrenaline content (- - □ - -) in guinea-pig atria after a single dose of reserpine (1 mg/kg). Values are means ± standard errors of the mean for six animals and are expressed as percentages of control values obtained from normal animals (control values for eight animals ± standard errors indicated by thin horizontal lines). (From Muskus, 1964.)

to catecholamines or their precursors occurs at a time when the accumulation of noradrenaline or adrenaline in the tissue is so small as to be barely detectable (Muscholl, 1960; Furchgott, Kirpekar, Rieker & Schwab, 1963; Trendelenburg & Crout, 1964). Crout *et al.* (1962), for example, found that the response of reserpinized guinea-pig atria to tyramine could be restored to 70% of normal values by exposure to noradrenaline, although the amount of noradrenaline accumulating in the tissue was only 2·2% of the normal level. These experiments lend further weight to the view that only a very small compartment of noradrenaline is necessary for normal adrenergic function. The refilling experiments, however, contain some puzzling observations. For instance, the responses to tyramine and nerve stimulation are not restored together; the restoration of adrenergic transmission is more difficult to achieve than the restoration of the tyramine response, but the effect is more prolonged (Trendelenburg & Pfeffer, 1964). The restoration of adrenergic transmission also seems to occur rather rapidly after the administration of noradrenaline precursors, suggesting an unexpectedly rapid synthesis of noradrenaline. Restoration or enhancement of adrenergic transmission can also be achieved by the administration of agents which cannot give rise to noradrenaline, and which normally lack pharmacological activity or have only weak activity at adrenergic receptors. Thus adrenergic transmission in the reserpinized nictitating membrane was enhanced by the administration of serotonin or by metanephrine and normetanephrine (Mirkin & Euler, 1963). Luduena & Snyder (1963) also found that D-noradrenaline was effective in restoring tyramine responsiveness in the reserpinized nictitating membrane, although this compound has much less pharmacological activity than L-noradrenaline.

5. Reserpine-resistant pools

As discussed in the previous chapter, the rate of depletion of noradrenaline from sympathetic nerves induced by reserpine is dependent, in many tissues, on the impulse flow in such nerves. Thus sympathetic decentralization, acute postganglionic denervation or ganglionic blocking drugs retard the disappearance of noradrenaline after reserpine (Holzbauer & Vogt, 1956; Kärki, Paasonen & Vanharkatano, 1959; Mirkin, 1961; Weiner, Perkins & Sidman, 1962; Hertting, Potter & Axelrod, 1962; Benmiloud & Euler, 1963). After preganglionic denervation of the sympathetic innervation of the cat gastrocnemius muscle it was reported that noradrenaline disappeared from the tissue in response to reserpine in two phases (Sedvall, 1964). Within the first 5 h after drug treatment some 85 % of the catecholamine content disappeared. The remaining 15 %, however, appeared to be resistant to depletion by reserpine and disappeared only slowly during the next 19 h (Fig. 9.6). The second slower phase of disappearance could also be detected in normally innervated muscles, but in this case was very much smaller (Fig. 9.6). The resistant pool of noradrenaline in the decentralized muscles could not be depleted by the administration of a second dose of reserpine. However, the reserpine-resistant fraction of noradrenaline could be rapidly depleted by electrical stimulation of the decentralized adrenergic nerves (Sedvall & Thorson, 1963). Sedvall (1964) suggested that this result implied the existence of two

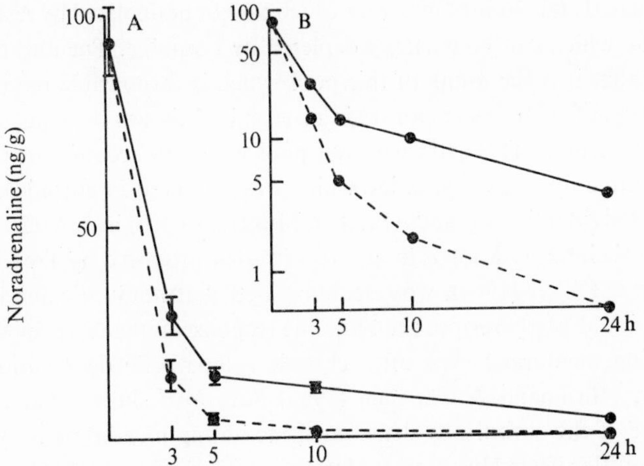

Fig. 9.6. Reserpine-resistant store of noradrenaline in skeletal muscle. Noradrenaline content (ng/g) of innervated (dotted line) and sympathetically decentralized skeletal muscles in the urethane anaesthetized cat at different times after reserpine (5 mg/kg, i.v.). In A and B the same results are presented but in B the noradrenaline content is plotted on a log. scale. Each point is the mean of at least six animals ± standard error of the mean. (From Sedvall, 1964.)

storage forms of noradrenaline in sympathetic nerves; one large pool containing about 85 % of the total catecholamine was rapidly depleted by reserpine, the remaining 15 % was stored in a reserpine-resistant pool which could be released by nerve stimulation.

In the rat, after removal of the superior cervical ganglion, the reserpine-induced depletion of H^3-noradrenaline from the acutely denervated salivary gland is retarded. Tyramine administration released the labelled amine from both normal and acutely denervated glands, but did not affect the difference between the two (Fischer & Kopin, 1964).

These results confirm the existence of an intraneuronal reserpine-resistant storage compartment, which can be released by nerve impulses, and suggested that this compartment was unaffected by tyramine.

Fischer, Kopin & Axelrod (1965) provide further evidence which suggests the existence of a second, *extraneuronal* pool of noradrenaline which is resistant to the actions of reserpine. In the chronically denervated rat salivary gland, small amounts of H^3-noradrenaline are still accumulated and retained in a store which cannot be depleted by reserpine but which can be partially depleted by tyramine. The authors suggest that it is the filling of this pool which is responsible for the restoration of tyramine responses in reserpinized tissues on exposure to noradrenaline. This extraneuronal pool may also account for the considerable uptake of tyramine which was observed in chronically denervated rat salivary glands (Fischer, Musacchio, Kopin & Axelrod, 1964). This interpretation is in accord with that proposed by Trendelenburg & Pfeffer (1964), who demonstrated that noradrenaline infusions could partly restore the tyramine response in the reserpinized nictitating membrane even after chronic sympathetic denervation. Haefely, Hürlimann & Thoenen (1963) have also shown that the response of the chronically denervated nictitating membrane to tyramine is little affected by reserpine treatment. Fischer *et al.* (1965) consider the possibility that the extraneuronal store in the rat salivary gland may be accounted for by the presence of a few anomalous sympathetic postganglionic cell bodies, axons and terminals in the tissue. These would be unaffected by ganglionectomy and could possibly account for the uptake and retention of small amounts of exogenous catecholamines in chronically denervated structures. In accord with this hypothesis, it is known that in the vas deferens, where a large part of adrenergic innervation is of this anomalous type (i.e. 'floating ganglion cells') (Owman & Sjöstrand, 1965), the noradrenaline content is unaffected by chronic denervation and is remarkably resistant to reserpine depletion (Sjöstrand, 1962). The noradrenaline content of sympathetic ganglia is also more resistant to depletion by reserpine than that in nerve terminals, especially after preganglionic denervation (Kirpekar, Cervoni & Furchgott, 1962).

6. Integrated schemes

In recent years there have been many schemes proposed to illustrate increasingly complex notions of the intraneuronal 'compartmentalization' of noradrenaline storage (Trendelenburg, 1961; Furchgott *et al.* 1963; Potter & Axelrod, 1963; Brodie & Beaven, 1963; Costa & Brodie, 1964; Kopin, 1964; Axelrod, 1965). Each scheme naturally incorporates the personal bias of the author and that illustrated in Fig. 9.7 is no exception to this rule. Such models can only be useful if regarded as a stimulus to further experiment, all are constructed in the certainty that next year will bring a better model. The illustration

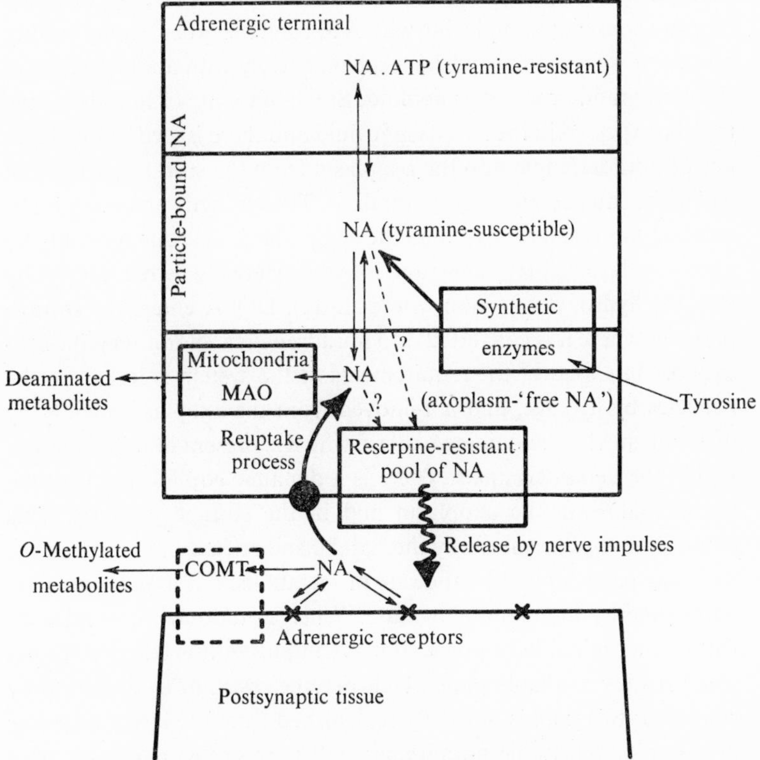

Fig. 9.7. Schematic representation of noradrenaline stores in an adrenergic terminal. For explanation see text.

presented in Fig. 9.7 is a synthesis of ideas previously expressed by many other authors.

Noradrenaline in sympathetic nerve terminals is stored mainly in particles, although the exact proportion of the total amine content which is particle-bound remains obscure. Noradrenaline in the particles may be present in more than one storage form, one of which is firmly bound by complexing with ATP, and the other is more readily releasable by tyramine and can exchange freely with noradrenaline in the extragranular axoplasmic fluid. Small amounts of noradrenaline may be expected in the axoplasmic fluid and there is a continual leakage of noradrenaline into the axoplasm from the storage particles in which the catecholamine is synthesized. The synthetic precursor is circulating L-tyrosine. This enters the sympathetic neurone (possibly by a mediated transport system) and is hydroxylated to form L-DOPA by tyrosine hydroxylase (unknown location). DOPA enters the storage particles where it is converted into noradrenaline. Noradrenaline synthesized in excess of the requirements of the system is lost from the particles by leakage, and is removed from the axoplasm mainly by metabolism via monoamine oxidase which is present in mitochondria within the nerve terminals. There is a dynamic equilibrium between noradrenaline in the axoplasm and in the storage particles. This equilibrium is determined by the uptake and release of noradrenaline from the particles and by the rate of metabolism of axoplasmic noradrenaline by monoamine oxidase. When noradrenaline is released from the terminal in response to nerve impulses, it is released from a small readily available pool which may be resistant to depletion by reserpine; this pool is normally replenished from the larger reservoir of reserpine-releasable noradrenaline. The reserpine-releasable pool can also be released by tyramine, although one part of this store is preferentially released by tyramine. The small pool of immediately available noradrenaline cannot be released by tyramine, but can possibly be released by bretylium. A second small pool of noradrenaline, which is resistant to reserpine depletion but cannot be released by nerve impulses, may be located outside the adrenergic terminal. This extraneuronal reserpine-resistant pool and the intraneuronal reserpine-resistant pool can be replenished by an uptake of exogenous catecholamines.

The uptake process (Uptake$_1$) which delivers extracellular noradrenaline into the axoplasm of the adrenergic terminal is located in the axonal membrane. From the axoplasm the accumulated catecholamine is rapidly distributed into other intraneuronal pools by a rapid uptake into storage particles. Noradrenaline released into the synaptic cleft in response to nerve impulses is mainly inactivated by a transport back into the nerve terminal in this way. Small amounts of the released catecholamine, however, overflow into the circulation or are metabolized by catechol-*O*-methyl transferase, which is probably located in postsynaptic tissues in the vicinity of the adrenergic receptors

It is interesting to note the similarities between this scheme for an adrenergic synapse with that described for cholinergic synapses by Birks & Mackintosh (1961). These authors proposed that in the cholinergic terminals of the cat superior cervical ganglion the following pools of acetylcholine could be distinguished:

(*a*) *Stationary A.Ch.* About 15 % of the total transmitter content of the ganglion is thought to be present in preterminal axons, from which it cannot be released by nerve impulses.

(*b*) A large part of the remaining 85% is present in synaptic regions as *depot A.Ch.* from which it can be released by nerve stimulation, although one part of this store may be more readily released by nerve impulses than another. Depot acetylcholine is thought to be present in storage vesicles in the nerve terminals.

(*c*) A small pool of *surplus A.Ch.* is present in the axoplasm, this pool is rapidly destroyed by acetylcholinesterase.

Acetylcholine liberated into the synaptic cleft in response to nerve impulses in the presynaptic terminal is rapidly inactivated by hydrolysis, catalysed by the enzyme acetylcholinesterase, which is located in the postsynaptic cell in the vicinity of the cholinergic receptors. The choline thus produced is recaptured by an uptake mechanism in the presynaptic terminal and can be re-used after resynthesis to acetylcholine.

Thus in both cholinergic and adrenergic terminals various compartments of transmitter can be distinguished, and in both cases the main stores of transmitter are probably in storage vesicles within the terminal. The inactivation of the released transmitter is in one case metabolic and in the other accomplished by a physical removal of the transmitter from its site of action. It may be that the requirement for a very rapid inactivation system in cholinergic synapses, which may operate at very high frequencies, can only be met satisfactorily by a rapid metabolism of the transmitter in the presence of a large excess of a degradative enzyme. In adrenergic synapses, however, which operate at very low frequencies, the removal of the transmitter by re-uptake into the presynaptic terminal is sufficiently rapid to meet the demands of the system without destroying the transmitter substance. In both cases there is a 'transmitter economy' in that either the transmitter itself or some important part of the molecule is recovered by the presynaptic terminal for subsequent re-use.

REFERENCES

Andén, N. E. (1964). Uptake and release of dextro- and laevo-adrenaline in noradrenaline stores. *Acta pharmac. tox.* **21**, 59–75.

Andén, N. E. & Magnusson, T. (1965). Functional significance of noradrenaline depletion by α-methyl metatyrosine, metaraminol and dextro-adrenaline. In *Pharmacology of Cholinergic and Adrenergic Transmission* (ed. Koelle, Douglas and Carlsson), pp. 319–328. Oxford: Pergamon Press.

Andén, N. E., Magnusson, T. & Waldeck, B. (1964). Correlation between noradrenaline uptake and adrenergic nerve function after reserpine treatment. *Life Sciences* **3**, 19–25.

Axelrod, J. (1965). The metabolism, storage and release of catecholamines. *Recent Prog. Horm. Res.* **21**, 597–622.

Axelrod, J., Gordon, E. K., Hertting, G., Kopin, I. J. & Potter, L. T. (1962). On the mechanism of tachyphylaxis to tyramine in the isolated rat heart. *Br. J. Pharmac. Chemother.* **19**, 56–63.

Benmiloud, M. & Euler, U. S. von (1963). Effects of bretylium, reserpine, guanethidine and sympathetic denervation on the noradrenaline content of the rat submaxillary gland. *Acta physiol. scand.* **59**, 34–42.

Birks, R. & Macintosh, F. C. (1961). Acetylcholine metabolism of a sympathetic ganglion. *Can. J. Biochem. Physiol.* **39**, 787–827.

Brodie, B. B. & Beaven, M. A. (1963). Neurochemical transducer systems. *Med. exp.* **8**, 320–351.

Burn, J. H. & Rand, M. J. (1958). The actions of sympathomimetic amines in animals treated with reserpine. *J. Physiol., Lond.* **144**, 314–336.

Burn, J. H. & Rand, M. J. (1960). The effect of precursors of noradrenaline on the response to tyramine and sympathetic stimulation. *Br. J. Pharmac. Chemother.* **15**, 47–55.

Campos, H. A. & Shideman, F. E. (1962). Subcellular distribution of catecholamines in the dog heart. Effect of reserpine and norepinephrine administration. *Int. J. Neuropharmac.* **1**, 13–22.

REFERENCES

Campos, H. A., Stitzel, R. E. & Shideman, F. E. (1963). Action of tyramine and cocaine on catecholamine levels in subcellular fractions of the isolated cat heart. *J. Pharmac. exp. Ther.* **141**, 290–300.

Carlsson, A. (1964). Functional significance of drug-induced changes in brain monoamine levels. *Prog. Brain Res.* **8**, 9–27.

Chidsey, C. A. & Harrison, D. C. (1963). Studies on the distribution of exogenous norepinephrine in the sympathetic neurotransmitter stores. *J. Pharmac. exp. Ther.* **140**, 217–223.

Costa, E., Boullin, D. J., Hammer, W., Vogel, W. & Brodie, B. B. (1966). Interactions of drugs with adrenergic neurones. (Second International Catecholamine Symposium.) *Pharmac. Rev.* **18**, 577–598.

Costa, E. & Brodie, B. B. (1964). Concept of the neurochemical transducer as an organized molecular unit at sympathetic nerve endings. *Prog. Brain Res.* **8**, 168–185.

Crout, J. R. (1964). The uptake and release of H^3-norepinephrine by the guinea pig heart *in vivo*. *Arch. exp. Path. Pharmak.* **248**, 85–98.

Crout, J. R., Muskus, A. J. & Trendelenburg, U. (1962). Effect of tyramine on isolated guinea-pig atria in relation to their noradrenaline stores. *Br. J. Pharmac. Chemother.* **18**, 600–611.

Dengler, H. J., Spiegel, H. E. & Titus, E. O. (1961). Uptake of tritium labelled norepinephrine in brain and other tissues of cat *in vitro*. *Science, N.Y.* **133**, 1072–1073.

Euler, U. S. von & Lishajko, F. (1965). Free and bound noradrenaline in the rabbit heart. *Nature, Lond.* **205**, 179–180.

Euler, U. S. von, Stjärne, L. & Lishajko, F. (1963). Uptake of radioactively labelled DL-catecholamines in isolated adrenergic nerve granules with and without reserpine. *Life Sciences* **2**, 878–885.

Fischer, J. E. & Kopin, I. J. (1964). Evidence for intra- and extraneuronal reserpine-resistant norepinephrine pools. *Fedn. Proc.* **23**, 349.

Fischer, J. E., Kopin, I. J. & Axelrod, J. (1965). Evidence for extraneuronal binding of norepinephrine. *J. Pharmac. exp. Ther.* **147**, 181–185.

Fischer, J. E., Musacchio, J., Kopin, I. J. & Axelrod, J. (1964). Effects of denervation on the uptake and β-hydroxylation of tyramine in the rat salivary gland. *Life Sciences* **3**, 413–420.

Furchgott, R. F., Kirpekar, S. M., Rieker, M. & Schwab, A. (1963). Actions and interactions of norepinephrine, tyramine and cocaine on aortic strips of rabbit and left atria of guinea pig and cat. *J. Pharmac. exp. Ther.* **142**, 39–58.

Gillespie, J. S. & Mackenna, B. R. (1961). The inhibitory action of the sympathetic nerves on the smooth muscle of the rabbit gut, its reversal by reserpine and restoration by catecholamines and by dopa. *J. Physiol., Lond.* **156**, 17–34.

Haefely, W., Hürlimann, A. & Thoenen, H. (1963). The response to tyramine of the normal and denervated nictitating membrane of the cat: analysis of the mechanism and sites of action. *Br. J. Pharmac. Chemother.* **21**, 27–38.

Hertting, G., Potter, L. T. & Axelrod, J. (1962). Effect of decentralization and ganglionic blocking agents on the spontaneous release of H^3-norepinephrine. *J. Pharmac. exp. Ther.* **136**, 289–292.

Holzbauer, M. & Vogt, M. (1956). Depression by reserpine of the noradrenaline concentration in hypothalamus of the cat. *J. Neurochem.* **1**, 8–11.

Iversen, L. L. (1963). The uptake of noradrenaline by the isolated perfused rat heart. *Br. J. Pharmac. Chemother.* **21**, 523–537.

Iversen, L. L., Glowinski, J. & Axelrod, J. (1965). Uptake and storage of H^3-norepinephrine in reserpine-pretreated rat heart. *J. Pharmac. exp. Ther.* **150**, 173–183.

Iversen, L. L. & Whitby, L. G. (1963). The subcellular distribution of catecholamines in normal and tyramine-depleted mouse hearts. *Biochem. Pharmac.* **12**, 582–584.

Kärki, N. T., Paasonen, M. K. & Vanhakartano, P. A. (1959). The influence of pentolonium, isoraunescine and yohimbine on the noradrenaline depleting action of reserpine. *Acta pharmac. tox.* **16**, 13–19.

Kirpekar, S. M., Cervoni, P. & Furchgott, R. F. (1962). Catecholamine content of the cat nictitating membrane following procedures sensitizing it to norepinephrine. *J. Pharmac. exp. Ther.* **135**, 180–190.

Kopin, I. J. (1964). Storage and metabolism of catecholamines: the role of monoamine oxidase. *Pharmac. Rev.* **16**, 179–191.

Kopin, I. J. & Bridgers, W. (1963). Differences in D- and L-norepinephrine. *Life Sciences* **2**, 356–362.

Kuschinsky, G., Lindmar, R., Lüllmann, H. & Muscholl, E. (1960). Der Einfluss von Reserpin auf die Wirkung der 'Neuro-sympathomimetica'. *Arch exp. Path. Pharmak.* **240**, 242–252.

Luduena, F. P. & Snyder, A. L. (1963). Reserpine, dextro-norepinephrine (NE) and tyramine interactions. *Biochem. Pharmac.* **12**, Suppl. p. 66.

Maickel, R. P., Beaven, M. A. & Brodie, B. B. (1963). Implications of uptake and storage of norepinephrine by sympathetic nerve endings. *Life Sciences* **2**, 953–958.

Mirkin, B. L. (1961). The effect of synaptic blocking agents on reserpine induced alterations in adrenal medullary and urinary catecholamine levels. *J. Pharmac. exp. Ther.* **133**, 34–40.

Mirkin, B. L. & Euler, U. S. von (1963). Effect of catecholamines on the responses of the nictitating membrane to nervous stimulation in cats treated with reserpine. *J. Physiol., Lond.* **168**, 804–819.

Musacchio, J., Kopin, I. J. & Weise, V. K. (1965). Subcellular distribution of some sympathomimetic amines and their β-hydroxylated derivatives in the rat heart. *J. Pharmac. exp. Ther.* **148**, 22–28.

Muscholl, E. (1960). Die Hemmung der Noradrenalin-Aufnahme des Herzen durch Reserpin und die Wirung von Tyramin. *Arch exp. Path. Pharmak.* **240**, 234–241.

Muskus, A. J. (1964). Evidence for different sites of action of reserpine and guanethidine. *Arch. exp. Path. Pharmak.* **248**, 498–513.

Neff, N. H., Tozer, T. N., Hammer, W. & Brodie, B. B. (1965). Kinetics of release of norepinephrine by tyramine. *Life Sciences* **4**, 1869–1875.

Norberg, K.-A., & Hamberger, B. (1964). The sympathetic adrenergic neurone. *Acta physiol. scand.* **63**, Suppl. 238.

Owman, Ch. & Sjöstrand, N. O. (1965). Short adrenergic neurones and catecholamine-containing cells in vas deferens and accessory male genital glands of different mammals. *Z. Zellforsch. mikrosk. Anat.* **66**, 300–320.

Potter, L. T. & Axelrod, J. (1963). Studies on the storage of norepinephrine and the effects of drugs. *J. Pharmac. exp. Ther.* **140**, 199–206.

Potter, L. T., Axelrod, J. & Kopin, I. J. (1962). Differential binding and release of norepinephrine and tachyphylaxis. *Biochem Pharmac.* **11**, 254–256.

Schümann, H. J., Schnell, K. & Philippu, A. (1964). Subcelluläre Verteilung von Noradrenalin und Adrenalin in Meerschweinchenherzen. *Arch. exp. Path. Pharmak.* **249**, 251–266.

Sedvall, G. (1964). Short term effects of reserpine on noradrenaline levels in skeletal muscle. *Acta physiol. scand.* **62**, 101–109.

Sedvall, G. & Thorson, J. (1963). The effect of nerve stimulation on the release of noradrenaline from a reserpine-resistant transmitter pool in skeletal muscle. *Biochem. Pharmac.* **12** (Suppl.), pp. 65–66.

Sjöstrand, N. O. (1962). Effect of reserpine and hypogastric denervation on the noradrenaline content of the vas deferens of the guinea pig. *Acta physiol. scand.* **56**, 376–380.

Stjärne, L. (1964). Studies of catecholamine uptake, storage and release mechanisms. *Acta physiol. scand.* **62**, Suppl. 228.

Trendelenburg, U. (1961). Modification of the effect of tyramine by various agents and procedures. *J. Pharmac. exp. Ther.* **134**, 8–17.

Trendelenburg, U. (1963). Supersensitivity and subsensitivity to sympathomimetic amines. *Pharmac. Rev.* **15**, 225–276.

Trendelenburg, U. & Crout, J. R. (1964). The norepinephrine stores of isolated atria of guinea-pigs treated with reserpine. *J. Pharmac. exp. Ther.* **145**, 151–161.

Trendelenburg, U. & Pfeffer, R. I. (1964). Effect of infusions of sympathomimetic amines on the response of spinal cats to tyramine and to sympathetic stimulation. *Arch. exp. Path. Pharmak.* **248**, 39–53.

Wegmann, A. & Kako, K. (1961). Particle-bound and free catecholamines in dog hearts and the uptake of injected norepinephrine. *Nature, Lond.* **192**, 978.

Weiner, N., Perkins, M. & Sidman, R. L. (1962). Effect of reserpine on noradrenaline content of innervated and denervated brown adipose tissue of the rat. *Nature, Lond.* **193**, 137–138.

10

CATECHOLAMINES IN THE CENTRAL NERVOUS SYSTEM

It is currently assumed that the majority of synapses in the central nervous system operate by the release of chemical transmitter substances from presynaptic nerve endings. However, because of the great anatomical complexity of brain tissue and the inaccessibility of central synapses the identification of the transmitter substances involved in central neurotransmission cannot yet be made with any certainty (Gaddum, 1963; Salmoiraghi, Costa & Bloom, 1965). In recent years the evidence that noradrenaline and dopamine may act as neurotransmitters in the central nervous system has become progressively more convincing. In this chapter only some aspects of the biochemical and pharmacological properties of central catecholamine-containing neurones will be reviewed. For a more detailed treatment of the functions and properties of catecholamine-containing neurones the reader is referred to the recent reviews by Fuxe, Dahlström & Hillarp (1965) and Glowinski & Baldessarini (1966).

1. Evidence for the existence of catecholamine-containing neurones in the central nervous system

Noradrenaline is found in appreciable amounts in mammalian brain tissue. Vogt (1954) showed that this amine is unequally distributed among various regions of the cat brain, the highest concentrations being found in the hypothalamus. The unequal distribution of noradrenaline among various brain regions suggested that the catecholamine might play some role in central adrenergic transmission, other than its possible role as the transmitter at vasomotor sympathetic nerve endings on blood vessels in the brain (Vogt, 1954, 1957). The regional distribution of noradrenaline has been examined in other mammalian species, including man, and found to follow the same general pattern as that first reported by Vogt (Bertler & Rosengren, 1959*a*; McGeer, Wada & McGeer, 1963; Sano *et al.* 1959) (Table 4.2).

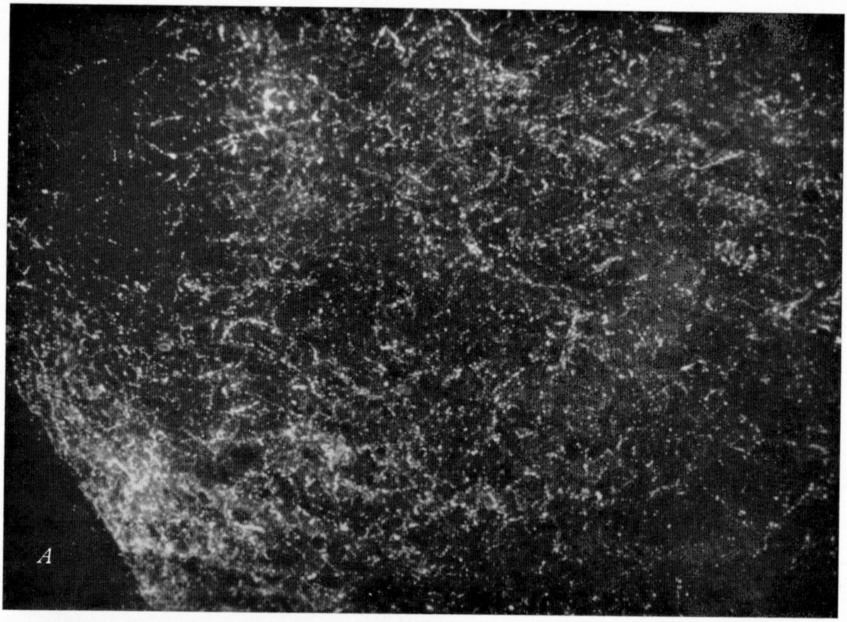

(*a*)

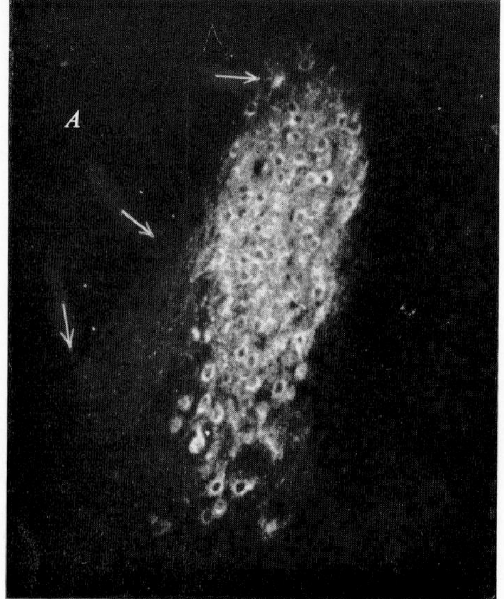

(*b*)

8 Fluorescence micrographs of brain.

(*a*) Transverse section of cat brain, ventral portion of nucleus dorsomedialis hypothalami, the third ventricle is seen to the left (A). A high density of catecholamine terminals is seen. Magnification ×100. From Fuxe (1965*a*).

(*b*) Transverse section of rat brain. A densely packed group of medium-sized strongly fluorescent catecholamine-containing nerve cell bodies (group A5) can be seen within the locus coeruleus. The cell mass is present just under the lateral surface of the fourth ventricle (*A*)—indicated by arrows. Magnification ×100. From Dahlström & Fuxe (1964*a*).

Dopamine is also present in the mammalian brain in amounts comparable to or exceeding noradrenaline (Montagu, 1957; Carlsson, Lindqvist, Magnusson & Waldeck, 1958; Carlsson, 1959). Dopamine has an entirely different regional distribution from that of noradrenaline, the highest concentrations being found in the corpus striatum of both animals and man (Bertler & Rosengren, 1959*a*; Sano *et al.* 1959; Ehringer & Hornykiewicz, 1960; Bertler, 1961).

The distributions of noradrenaline and dopamine in the bird brain have also been reported (Pscheidt & Haber, 1965; Aprison & Takahashi, 1965). Noradrenaline is also found in the brains of lower vertebrates, including reptiles and fishes. It is interesting to note that in the amphibian species examined the predominant catecholamine found in the brain is adrenaline rather than noradrenaline (Brodie, Bogdanski & Bonomi, 1964).

Application of the fluorescence histochemical technique for the cellular localization of monoamines to the brain has led to rapid advances in our knowledge of the precise localization of catecholamines in the central nervous system. It is now clear that noradrenaline and dopamine are concentrated in complex systems of catecholamine-containing neurones in the brain. In many areas of the brain fine nerve fibres containing high concentrations of catecholamine and with characteristic varicosities along their length have been described (Carlsson, Falck & Hillarp, 1962; Carlsson, Falck, Fuxe & Hillarp, 1964; Dahlström & Fuxe, 1964*a*, 1965; Fuxe, 1964, 1965*a*,*b*). In certain brain areas, notably in the lower brain stem, groups of nerve cell bodies are found which contain low concentrations of catecholamine in their cytoplasm (Dahlström & Fuxe, 1964*a*) (Plate 8). The use of the histochemical fluorescence technique in combination with specific lesions in the central nervous system has shown the disappearance of catecholamine-containing terminals from various parts of the brain or spinal cord; at the same time retrograde changes in the catecholamine fluorescence of cell bodies in the medulla oblongata and mesencephalon have been observed (Andén, Carlsson, Dahlström, Fuxe, Hillarp & Larsson, 1964; Andén, Häggendal, Magnusson & Rosengren, 1964; Carlsson *et al.* 1964; Dahlström & Fuxe, 1965). These changes are interpreted to suggest that the catecholamine-containing terminals observed in many parts of the brain and spinal cord represent the terminal branches of nerve fibres which arise from the cell bodies of catecholamine-containing neurones in the brain stem or—in the case of dopamine terminals in the caudate nucleus—in the substantia nigra. Other studies, however, which have shown a selective loss of noradrenaline from certain forebrain regions after lesions of the medial forebrain bundle and dorsomedial and ventromedial tegmentum have been interpreted differently (Heller & Moore, 1965). These authors suggest that the loss of catecholamine from forebrain regions is due, at least in part, to trans-synaptic effects, and point out that the neurochemical consequences of brain lesions may be complex and not explicable on the basis of the simple denervation studies performed on peripheral nerves.

The use of lesions, however, will certainly prove to be of great value in mapping the precise anatomical connections of catecholamine-containing neurones in the central nervous system. The preterminal axons of such neurones contain too little catecholamine to allow their demonstration by the fluorescence histochemical technique in normal tissue. However, after axotomy increased amine levels rapidly accumulate in the proximal part of the axon, especially at the region just proximal to the lesion. The increased fluorescence observed in neurone cell bodies and preterminal axons after axotomy is thus a most useful tool for mapping the connections of central catecholamine neurones (Dahlström & Fuxe, 1964*b*, *c*).

The fluorescent histochemical studies leave little doubt that specific systems of catecholamine-containing neurones exist in the central nervous system. The catecholamine neurones in the brain are similar in many respects to sympathetic postganglionic neurones in the peripheral nervous system. The two types of neurone are approximately similar in size and in the characteristic varicose appearance of their very fine terminals. The catecholamine content of central neurones is unevenly distributed throughout the cell, as in peripheral adrenergic neurones. Only low concentrations of catecholamine are observed in the cell bodies and preterminal axons, while extremely high concentrations are found in the terminal fibres, where the highest concentrations are present in the varicosities. Catecholamine-containing terminals in the brain have been observed to make contact with dendrites or cell bodies of other neurones, which may or may not contain catecholamines (Fuxe, 1965*a*, *b*).

Unfortunately, it has not been possible to make an absolutely clear distinction between dopamine- and noradrenaline-containing neurones, since the fluorescent products of these two amines have almost identical fluorescence characteristics. The fluorescence histochemical studies have relied on pharmacological tests in order to distinguish between dopamine- and noradrenaline-containing neurones (Fuxe, 1965*a*; Carlsson, Dahlström, Füxe & Hillarp, 1965). It may be hoped that improvements in the specificity of the histochemical technique will allow a more precise distinction between dopamine and noradrenaline neurones. On the basis of present evidence, it is suggested that while the noradrenaline-containing terminals have a widespread distribution from the spinal cord to the neocortex, dopamine-containing terminals are mainly found only in circumscribed regions of the telencephalon, i.e. in the neostriatum, n.accumbens and the tuberculum olfactorium. Noradrenaline terminals are particularly abundant in the hypothalamus and certain visceral afferent and efferent nuclei of the cranial nerves (Fuxe, 1965*a*, *b*).

The existence of noradrenaline and dopamine in separate systems of neurones in the brain and the high concentrations of these amines which are found in the terminals of these neurones constitute strong suggestive evidence that these amines act as neurotransmitters in the central nervous system. The evidence in favour of noradrenaline playing a role in neurotransmission in the central nervous system is further strengthened by the anatomical similarities between central noradrenaline-containing neurones and sympathetic postganglionic neurones, in which this amine has been positively identified as the neurotransmitter.

2. Metabolism of brain catecholamines

(*a*) *Biosynthesis*

In common with other neurotransmitter substances noradrenaline and dopamine should be synthesized intraneuronally in the central catecholamine-containing neurones. All the enzymes necessary for the synthesis of dopamine and norepinephrine from tyrosine have been detected in mammalian brain tissue: tyrosine hydroxylase is present in the brain and has been studied in the bovine caudate nucleus (Bagchi & McGeer, 1964). The enzyme from this source has properties similar to those described for the enzyme in the peripheral sympathetic nervous system or in the adrenal medulla (chapter 5). In homogenates of the bovine caudate nucleus, tyrosine hydroxylase activity is specifically associated with a fraction of nerve-ending particles which can be isolated by density-gradient centrifugation and which contains considerable amounts of endogenous noradrenaline and dopamine (McGeer, Bagchi & McGeer, 1965), suggesting that the enzyme may be localized in the nerve terminals which contain endogenous catecholamines. Aromatic-L-amino acid decarboxylase is widely distributed among various brain areas, in amounts approximately proportional to the catecholamine content of the various regions (Bertler & Rosengren, 1959*b*). The activity of this enzyme in the spinal cord is decreased in regions caudal to a total transection, at a time when the catecholamine content of catecholamine-containing terminals in these regions is also diminished (Andén, Magnusson & Rosengren, 1965). This again suggests that the enzyme is present within catecholamine-containing terminals in the central nervous system. However, the situation is complicated by the fact that a single enzyme appears to be responsible for the formation of both dopamine and serotonin in the brain and other tissues. The existence of separate serotonin-containing neurones in the central nervous system (Fuxe, 1964, 1965*a*, *b*) means that this enzyme is present not only in catecholamine-containing neurones but also in serotonin-containing neurones. This may explain why the regional distribution of the enzyme does not exactly parallel the regional distribution of catecholamines in the brain and spinal cord. Dopamine-β-hydroxylase is also present in the brain, in amounts approximately correlating with the catecholamine content of the various brain regions (Udenfriend & Creveling, 1959). While the brain may be able to synthesize small amounts of adrenaline (McGeer & McGeer, 1964; Milhaud & Glowinski, 1963), there is little evidence that this amine accounts for any appreciable proportion of the catecholamine content of the mammalian brain.

Because of the failure of dopamine to penetrate the blood–brain barrier, most experiments on catecholamine synthesis *in vivo* have made use of parenteral injections of radioactively labelled or non-labelled DOPA. When large amounts of DOPA are administered in this way, high concentrations of dopamine accumulate in various

Table 10.1. *Formation of noradrenaline from DOPA and from dopamine in the rat brain*

(From Glowinski & Iversen, 1966*a*.)

Brain region	After H^3-dopamine			After H^3-DOPA		
	H^3-dopamine	H^3-NA	Dopamine (%)	H^3-dopamine	H^3-NA	Dopamine (%)
Hypothalamus	91	168	35	2191*	177*	92*
Medulla oblongata	32	83	28	2689	341	89
Cerebellum	14	42	25	1249	399	76
Striatum	156	25	86	1574	115	93

* Five hours after H^3-DOPA injection.

Brain regions were assayed for H^3-dopamine and H^3-noradrenaline 2 h after the injection of H^3-dopamine (1·7 μg) or H^3-DOPA (1·7 μg) into the lateral ventricle. Results are mean values for groups of 5–10 experiments, values are expressed in mμc. Dopamine percentage = dopamine as percentage of dopamine + NA.

brain regions, but rather little noradrenaline appears to be formed (Carlsson, 1964; Bertler & Rosengren, 1959*b*). The possibility that exogenous DOPA may be taken up and decarboxylated to form dopamine in both catecholamine-containing and serotonin-containing neurones should be remembered (Kuntzman, Shore, Bogdanski & Brodie, 1961). In experiments in which radioactively labelled dopamine was injected directly into the lateral ventricle of the rat brain we have found that the labelled catecholamine is rapidly converted into noradrenaline. If, however, labelled DOPA is injected in the same way, large amounts of dopamine are formed, which are not then rapidly further converted into noradrenaline (Glowinski & Iversen, 1966*a*). These results suggest that much of the dopamine formed in the brain from exogenous DOPA may be formed in cells which lack the ability to synthesize noradrenaline, i.e. it may be formed in structures which do not normally contain catecholamines (Table 10.1).

The rate of turnover of catecholamines in the brain has been estimated by following the decline in specific activity of brain catecholamines after the administration of labelled precursors such as tyrosine

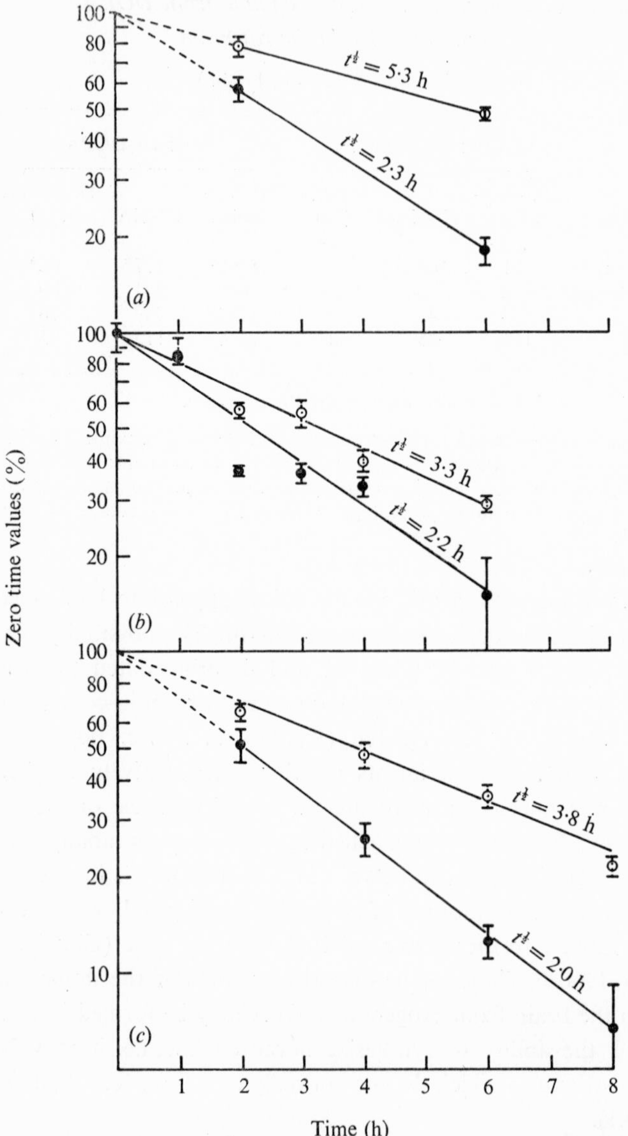

Fig. 10.1. Turnover of brain catecholamines measured by three methods. Rate of turnover of noradrenaline in hypothalamus and cerebellum of rat brain, ○, Hypothalamus; ●, cerebellum. In (*a*) turnover was measured by following the dis-

or DOPA. These studies have indicated that the brain catecholamines have a rapid rate of turnover, with half times of approximately 2 h for dopamine and 4 h for noradrenaline in mammalian brain (Udenfriend & Zaltzman-Nirenberg, 1963; Burack & Draskoćzy, 1964). In the rat brain we have examined the rate of turnover of catecholamines in various brain regions (Iversen & Glowinski, 1966). In these studies the rate of turnover of noradrenaline was measured by three alternative experimental approaches: (*a*) by measuring the rate of disappearance of endogenous catecholamine after the complete inhibition of catecholamine biosynthesis by a single large dose of the tyrosine hydroxylase inhibitor, α-methyl-*p*-tyrosine; (*b*) by measuring the decline in specific activity of noradrenaline in various regions after the injection of small amounts of labelled dopamine into the lateral ventricle; and (*c*) by measuring the decline in specific activity of noradrenaline in various brain regions after the intraventricular injection of small amounts of labelled noradrenaline. The results obtained by the three methods were very similar (Fig. 10.1). It was clear that considerable differences in the rate of turnover of noradrenaline existed among the various brain regions, the turnover being fastest in the cerebellum ($t\frac{1}{2} = 2$ h) and the cortex and hippocampus ($t\frac{1}{2} = 3$ h), and slower in the hypothalamus and medulla oblongata ($t\frac{1}{2} = 4$ h). These results are summarized in Table 10.2. Similar results have been reported by Fuxe (1966), who found that the fluorescent catecholamine-containing terminals of neocortical brain regions were depleted more rapidly than those in the brain stem or hypothalamus after the inhibition of catecholamine biosynthesis.

---

appearance of endogenously synthesized radioactive noradrenaline after the injection of tritiated dopamine into the lateral ventricle. In (*b*) the rate of disappearance of the endogenous noradrenaline content was measured after the blockade of noradrenaline biosynthesis with a single large dose of α-methyl-*p*-tyrosine. In (*c*) the rate of disappearance of H^3-noradrenaline was measured after the injection of the labelled catecholamine into the lateral ventricle. Half times for noradrenaline turnover in the two brain regions by the three methods are indicated on the figure. Similar results were obtained for other brain regions (Table 10.2). (From Iversen & Glowinski, 1966.)

Table 10.2. *Rate of turnover of noradrenaline in different regions of rat brain*

(Summary of results obtained by the three methods used to measure noradrenaline turnover in the rat brain (see Fig. 10.2). From Iversen & Glowinski, 1966.)

Brain region	Mean half time for all methods (h)	Upper and lower confidence limits (h)
Cerebellum	2·2*	1·86–2·60
Cortex	2·7	2·25–3·42
Hippocampus	3·0	2·42–3·84
Medulla oblongata	3·6	2·80–4·91
Hypothalamus	3·7	2·89–5·20

* Significantly different ($P < 0{\cdot}05$) from all other brain regions.

(*b*) *Metabolism*

The enzymes monoamine oxidase and catechol-*O*-methyl transferase are widely and almost uniformly distributed in various brain regions (Bogdanski, Weissbach & Udenfriend, 1957; Axelrod, Albers & Clemente, 1959). Until recently little was known of the metabolism of brain catecholamines in the intact brain. Mannarino, Kirshner & Nashold (1963) found that *O*-methylation played an important role in the metabolism of radioactively labelled noradrenaline infused into the lateral ventricles of the cat brain. The development of a technique of stereotaxic injection which allows the administration of small amounts of high specific activity H^3-noradrenaline into the lateral ventricle of the rat brain has greatly advanced studies of the metabolism of catecholamines in the brain *in vivo* (Milhaud & Glowinski, 1962, 1963). When small amounts of labelled noradrenaline are injected into the rat brain in this way, the major metabolites accumulating in the brain are *O*-methylated deaminated compounds (Table 10.3), suggesting that both *O*-methylation and deamination play an important role in the metabolism of brain catecholamines, as in the peripheral sympathetic nervous system (Glowinski, Kopin & Axelrod, 1965). The role of monoamine oxidase, however, is difficult to appraise, because of the important involvement of this enzyme in the metabolism of serotonin in the central nervous system. While the enzyme

Table 10.3. *Metabolism of* H^3*-noradrenaline in the rat brain*

(From Glowinski, Kopin & Axelrod, 1965.)

Metabolites	Percentage of total radioactivity in brain
Unchanged H^3-noradrenaline	55·7
H^3-Normetanephrine	12·9
H^3-Deaminated catechols	4·0
H^3-*O*-Methylated deaminated metabolites	27·1

Metabolites of H^3-noradrenaline in rat brain 1 h after the injection of 0·13 μg of the radioactive catecholamine into the lateral ventricle. H^3-Deaminated catechols = 3,4-dihydroxymandelic acid and the corresponding glycol; H^3-*O*-methylated deaminated metabolites were largely 3-methoxy-4-hydroxyphenylglycol and small amounts of VMA.

catechol-*O*-methyl transferase has been found to be associated with a nerve-ending fraction by density-gradient centrifugation studies, no such specific association of monoamine oxidase with nerve endings can be found (Alberici, De Lores Arnaiz & De Robertis, 1965; De Lores Arnaiz & De Robertis, 1962). There may also be species differences in the role of monoamine oxidase in the metabolism of brain catecholamines. Certain mammalian species such as the cat and dog fail to respond to monoamine oxidase inhibitors with the elevation of brain catecholamine levels observed in most other species (Vogt, 1964; Pscheidt, Morpurgo & Himwich, 1964). Histochemical studies have shown that monoamine oxidase inhibitors lead to a great increase in the fluorescence of serotonin-containing neurones while the catecholamine-containing neurones are less affected (Dahlström & Fuxe, 1964*a*).

3. Uptake and storage of catecholamines in central neurones

(*a*) *Uptake of exogenous catecholamines*

Because of the inability of catecholamines to penetrate the blood–brain barrier (Weil-Malherbe, Axelrod & Whitby, 1961), studies of the uptake of exogenous catecholamines in the brain have been much more difficult than the similar studies which have been performed in peripheral sympathetic nerves. Dengler, Spiegel & Titus (1961) found that slices of cat cerebral cortex were able to accumulate labelled noradrenaline when incubated with the labelled amine *in vitro*. The uptake process in the brain slices was found to saturate with increasing concentrations of noradrenaline in the incubation medium, indicating a K_m for the uptake of DL-noradrenaline of approximately 4×10^{-7} M; a similar value has been reported by Ross & Renyi (1964) for the uptake of DL-noradrenaline by slices of mouse cerebral cortex. The K_m for noradrenaline uptake by mammalian brain tissue is thus closely similar to that found for the uptake of DL-noradrenaline into sympathetic nerves in the rat heart ($6{\cdot}7 \times 10^{-7}$ M). The uptake of noradrenaline into brain slices could also be inhibited by the same drugs which inhibit uptake in peripheral sympathetic nerves (cocaine, phenoxybenzamine, imipramine, desmethylimipramine, amphetamine and other sympathomimetic amines) (Dengler, Michaelson, Spiegel & Titus, 1962; Ross & Renyi, 1964). Hamberger & Masuoka (1965) have recently shown that the noradrenaline or α-methyl noradrenaline accumulated by slices of rat cerebral cortex or brain stem is specifically localized in the axons and terminals of catecholamine containing neurones, suggesting that central catecholamine neurones, in common with peripheral catecholamine neurones, have the ability to take up and retain exogenous catecholamines. The histochemical studies of Fuxe & Hillarp (1964) have also demonstrated that the catecholamine-containing neurones in brain areas lacking a blood–brain barrier can take up and retain noradrenaline and α-methyl noradrenaline after the peripheral administration of these amines to reserpine-nialamide treated rats.

Similarly, after the injection of small amounts of labelled noradrenaline into the lateral ventricle of the rat brain, the labelled amine is selectively accumulated in catecholamine-containing neurones in the brain (Glowinski & Axelrod, 1966; Reivitch & Glowinski, 1966; Glowinski & Iversen, 1966 *a*). In studies of the regional distribution of the labelled catecholamine in the rat brain after intraventricular injection we found that after the injection of labelled noradrenaline large amounts of the labelled amine could be found in the striatum, which contains only small amounts of endogenous noradrenaline but large amounts of endogenous dopamine, suggesting that exogenous catecholamines may be taken up and retained indiscriminately by noradrenaline and dopamine-containing neurones. After the intraventricular injection of labelled noradrenaline the amine disappears only slowly from the brain regions in which it is accumulated (Fig. 10.1), this method can therefore be used as a useful tool for the introduction of small amounts of radioactively labelled noradrenaline into the brain catecholamine stores. The remarkable ability of central catecholamine-containing neurones to accumulate exogenous catecholamines again suggests a similarity between the structure and properties of central and peripheral catecholamine neurones. The uptake of extraneuronal catecholamines into central catecholamine neurones may act as an inactivation process for the catecholamines released at central synapses, as at peripheral adrenergic synapses.

(*b*) *Subcellular distribution of endogenous and exogenous catecholamines in the brain*

As discussed in chapter 4, noradrenaline in homogenates of peripheral sympathetically innervated tissues is localized mainly in particles of microsomal dimensions. These particles probably correspond to synaptic vesicles or storage particles liberated from the sympathetic nerve terminals on homogenization. When the brain is homogenized, however, most of the nerve terminals are not disrupted but pinch-off to form discrete nerve-ending particles or 'synaptosomes' (Gray & Whittaker, 1962). A considerable proportion of the endogenous noradrenaline in brain homogenates can be recovered in a fraction of synaptosomes on density-gradient centrifugation (Potter & Axelrod, 1963; Chruściel, 1960; Laverty, Michaelson, Sharman & Whittaker, 1963; Levi & Maynert, 1964). In keeping with the view that intraventricularly injected H^3-noradrenaline is specifically accumulated in central catecholamine neurones, the H^3-noradrenaline in brain homogenates similarly has been found to be localized in synaptosomal fractions on density-gradient centrifugation (Glowinski, Snyder & Axelrod, 1966; Snyder, Glowinski & Axelrod, 1965; Glowinski & Iversen, 1966*b*) (Fig. 10.2). In the latter study, we examined the subcellular distribution of endogenous and exogenous catecholamines in various regions of the rat brain. Marked differences in the subcellular distribution patterns were found between brain regions. For example the particulate/supernatant ratios for both endogenous and exogenous noradrenaline were considerably higher in the hypothalamus and medulla oblongata than in the cerebellum or cortex (Table 10.4). The particle/supernatant ratio in brain tissue

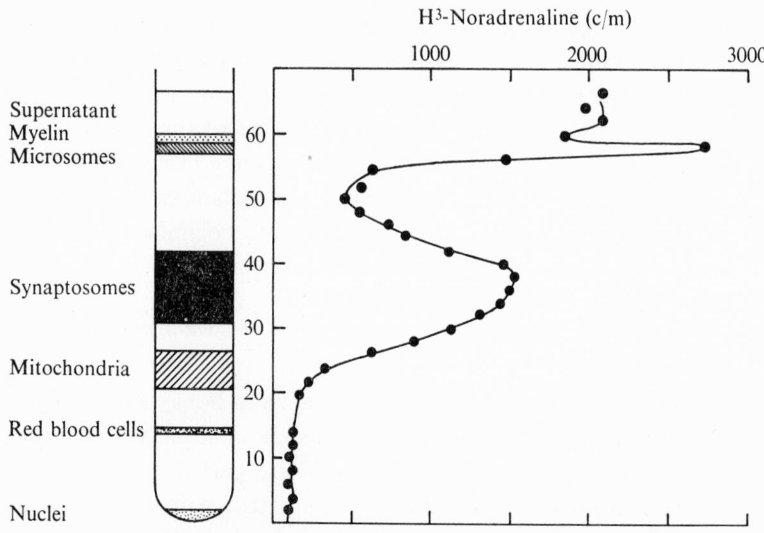

Fig. 10.2. Subcellular distribution of H^3-noradrenaline in rat brain. Distribution of H^3-noradrenaline on a sucrose density gradient after centrifugation of a sample of striatum homogenate obtained 4 h after the injection of H^3-noradrenaline (0·7 μg) into the lateral ventricle of the rat brain. Appearance of the gradient indicated on left. Five-drop samples were collected after puncturing the bottom of the tube, alternate fractions were assays for H^3-noradrenaline. Most of the H^3-noradrenaline is localized in a fraction of pinched-off nerve endings ('synaptosomes') and some is present in the supernatant fluid and in a microsomal particle fraction lying immediately below the supernatant layer. Similar results were obtained for other brain areas. (From Glowinski & Iversen, 1966*b*.)

should not be confused with the particle/supernatant ratio for catecholamines in peripheral tissues. In peripheral sympathetic nerves the particulate/supernatant ratio may indicate the ratio of noradrenaline stored in intraneuronal particles to that present as free noradrenaline in such nerves; in the brain, however, the particle fraction is much more complex, consisting of intact nerve terminals as well as storage particles liberated from the disruption of some nerve endings. The

Table 10.4. *Subcellular distribution of endogenous and exogenous noradrenaline in various regions of the rat brain*

(From Glowinski & Iversen, 1966*b*.)

Brain region	Total endogenous NA content (ng/g)	P/S ratio (endog. NA)	H^3-noradrenaline content (mμc/g)	P/S ratio (H^3-NA)
Cerebellum	143	1·54±0·15	177	1·16±0·50
Cortex	289	2·52±0·14	174	1·36±0·10
Medulla oblongata	564	3·35±0·10	606	2·51±0·16
Hypothalamus	1774	4·27±0·30	1177	2·83±0·26

One hour after the injection of H^3-noradrenaline into the lateral ventricle of the rat brain, regions were homogenized in isotonic sucrose and centrifuged for 1 h at 70000*g* to yield particulate (P) and supernatant (S) fractions which were then assayed for H^3-noradrenaline and for endogenous noradrenaline. Each value is the mean±S.E.M. for 5–6 determinations each involving pooled brain structures from two animals. The P/S ratios for either endogenous NA or for H^3-NA in any brain region are significantly different ($P < 0{\cdot}05$) from those in any other brain region.

supernatant fraction in the brain is also complex, arising partly from catecholamine liberated from disrupted nerve terminals and also from catecholamine liberated by the disruption of nerve cell bodies and preterminal axones.

Electron-microscopic investigations have revealed the presence of granulated vesicles in nerve endings of the anterior hypothalamus, which resemble those described in peripheral adrenergic terminals (De Iraldi, Duggan & De Robertis, 1963). The number of dense-core vesicles in the hypothalamic nerve terminals has been reported to vary in direct proportion to the catecholamine content of the hypothalamus being reduced by reserpine and other amine-depleting agents (Matsuoka, Ishii, Shimizu & Imaizumi, 1965). Small amounts of noradrenaline can be recovered in a microsomal particle fraction when catecholamine-containing synaptosome fractions are disrupted by hypotonic shock (Maynert, Levi & De Lorenzo, 1964; De Robertis, De Iraldi, De Lores Arnaiz & Zieher, 1965) but the evidence concerning the isolation of storage vesicles from central catecholamine-containing terminals in this way is controversial (Whittaker, 1965).

4. Some aspects of the pharmacology of brain catecholamines

The discovery that reserpine and many other centrally acting drugs can affect the storage and metabolism of catecholamines in the brain has stimulated great interest in the actions of drugs on brain catecholamines. The finding of precisely measurable chemical changes in the brain which appear to correlate well with changes in behaviour is of course very stimulating and intriguing, since it points a possible direction to one of the ultimate goals of neurochemistry: an understanding of the means by which changes in brain chemistry can influence the complex functions of the central nervous system. Many of the studies of the actions of drugs on peripheral sympathetic nerves in recent years have been partly motivated by the desire to understand the precise mode of action of centrally acting drugs. The similarities between catecholamine-neurones in the brain and postganglionic sympathetic neurones have already been pointed out—it may be a reasonable assumption that if the precise molecular actions of drugs were understood on the relatively simple and accessible peripheral adrenergic neurones it might be possible to extrapolate these findings to explain drug actions on central catecholamine neurones in similar or even identical terms. The evidence so far available has in general indicated that the mode of action of drugs on peripheral and central catecholamine neurones is indeed closely similar. However, much more will have to be known about the functions of catecholamine neurones in the central nervous system before it will be possible to predict the behavioural consequences of drug actions on these neurones.

In the present review only a small part of the now considerable literature on the pharmacology of brain catecholamines can be mentioned. Many studies have sought to establish a relation between the effects of drugs on the overall storage level of catecholamines in the brain with the effects of these drugs on behaviour (Brodie, Spector & Shore, 1959; Everett & Weigard, 1962; McGeer, Wada & McGeer, 1963; Carlsson, 1964). In general it has not been possible to obtain a good correlation between brain catecholamine levels and behavioural state, and indeed this is not surprising since brain amine levels can give only gross information on the nature of drug actions on brain catecholamines. As pointed out previously, the storage level of catecholamine in a tissue is not a static measure, but represents a dynamic equilibrium between the rate of formation and the rate of destruction of the catecholamine. Drugs which alter this steady state level may do so in several quite different ways (chapter 8).

(*a*) *Actions of reserpine*

The discovery that reserpine produces a long-lasting depletion of brain catecholamines (Holzbauer & Vogt, 1956; Carlsson, Rosengren, Bertler & Nilsson, 1957) was perhaps the single most important finding which catalysed the subsequent great interest in the actions of drugs on brain catecholamines. The actions of reserpine on brain catecholamines appear to be very similar to its actions on peripheral catecholamine stores. In both cases the drug causes an almost complete depletion of amine stores, which recover only after several weeks to normal levels. In the reserpinized brain exogenous noradrenaline is no longer accumulated in catecholamine neurones unless monoamine oxidase activity is inhibited (Fuxe & Hillarp, 1964; Glowinski, Iversen & Axelrod, 1966*a*). The increased metabolism of catecholamines by monoamine oxidase in the reserpinized brain is also reflected by accumulations of deaminated metabolites in the tissue (Roos & Werdinius, 1962), and is similar to the situation found in peripheral sympathetic nerves after reserpine (chapter 8). It is interesting that after reserpine treatment the recovery of catecholamine content is seen first in the cell bodies of the catecholamine-containing neurones; in these recovery may first be apparent in a ring of cytoplasm surrounding the cell nucleus (Dahlström & Fuxe, 1964*a*; Dahlström, Fuxe & Hillarp, 1965). This may suggest that the recovery from reserpine is prolonged because it depends on the formation of new storage particles in the cell bodies of catecholamine-containing neurones, which must then be transported to the nerve terminals. During chronic reserpine treatment it has been found that the small daily variations in the noradrenaline content of the severely depleted amine stores in the brain correlates well with variations in the animals' behaviour, suggesting that, as in peripheral adrenergic nerves, the presence of only very small amounts of noradrenaline are required for normal function (Häggendal & Lindqvist, 1964). Furthermore, the recovery from the pronounced sedative effects of a single dose of the drug occurs

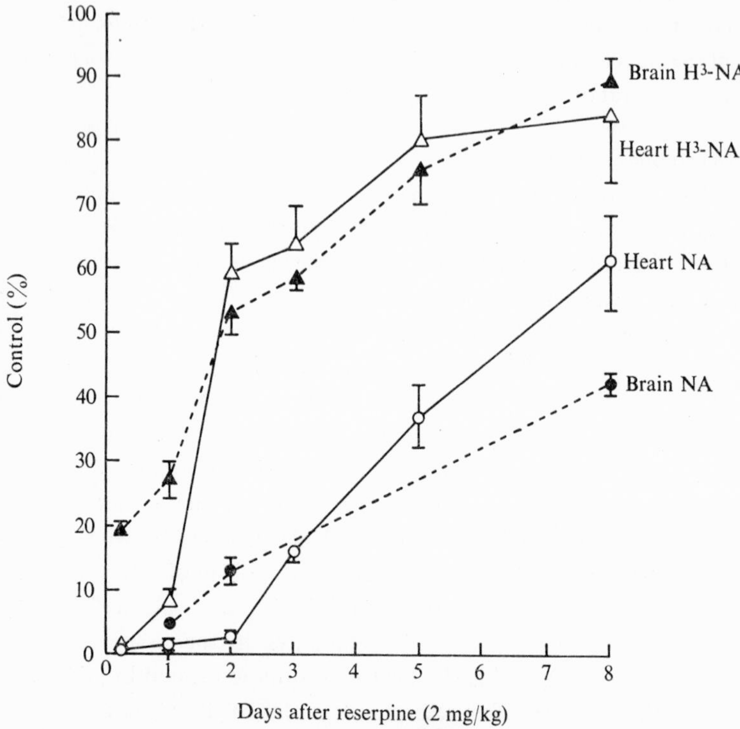

Fig. 10.3. Recovery of the ability to accumulate small amounts of H^3-noradrenaline (NA) and recovery of endogenous noradrenaline content in rat brain after a single dose of reserpine. Rats received an injection of H^3-noradrenaline into the lateral ventricle of the brain and the brain content of H^3-noradrenaline and of endogenous noradrenaline were assayed 1 h later. Each point is the mean ± standard error of the mean for 5–6 rats. The results are expressed as percentages of control values obtained in normal animals. Similar results for the rat heart (Fig. 9.4) are included for comparison. (From Glowinski, Iversen & Axelrod, 1966*a*.)

between 24 and 48 h after drug administration, at a time when the brain catecholamine levels remain severely depressed. As in the peripheral sympathetic nerves, however, a rapid recovery occurs at this time in the ability of the central catecholamine-containing neurones to take up and retain exogenous noradrenaline (Glowinski, Iversen & Axelrod, 1966*a*) (Fig. 10.3).

It should hardly be necessary to point out that the actions of reserpine on brain amines are not restricted to catecholamines. The drug also produces a severe and long-lasting depletion of brain serotonin stores, and it remains unclear whether the behavioural effects of the drug depend on the depletion of catecholamines or on the depletion of serotonin, or indeed whether either system is involved.

(*b*) *Inhibitors of catecholamine uptake*

If the ability of central catecholamine neurones to take up exogenous catecholamines represents an inactivation mechanism analogous to that described in peripheral sympathetic nerves, then drugs that inhibit this uptake process may also prevent the inactivation of catecholamines in the brain. As discussed above, it has been found that many of the drugs which inhibit the uptake of exogenous noradrenaline into peripheral sympathetic nerves also block the uptake of noradrenaline into brain slices incubated *in vitro*. Furthermore, it has recently been found that pretreatment with such drugs (imipramine, amitryptyline, amphetamine) can also inhibit the accumulation of labelled noradrenaline in central catecholamine neurones *in vivo*, after the intraventricular injection of the labelled amine (Glowinski & Axelrod, 1966; Glowinski, Iversen & Axelrod, 1966*b*). It might be expected that such drugs will have the property of potentiating the actions of catecholamines at central adrenergic synapses, in the same way as they potentiate the actions of catecholamines at peripheral adrenergic synapses. Such an action may have some relation to the antidepressant actions of drugs such as imipramine and amitryptyline (Glowinski & Axelrod, 1966) and to the central excitant actions of cocaine.

(*c*) *Amphetamine*

Amphetamine acts as a powerful central stimulant drug in most animals. There have been several biochemical and pharmacological hypotheses concerning the mode of action of this drug (Stein, 1964). One hypothesis is that amphetamine acts as an indirect sympathomimetic amine in the brain, i.e. it produces its central actions by a release of noradrenaline from central stores. This hypothesis is supported by the finding that large doses of D-amphetamine can cause a lowering in the levels of noradrenaline in the brain (Moore, 1963) and

Table 10.5. *Effects of drugs on the uptake and metabolism of H^3-noradrenaline in the rat brain*

(H^3-Noradrenaline and metabolites as percentages of values in normals. From Glowinski, Iversen & Axelrod, 1966*b*.)

Brain region	NA	NMN	Catechol deaminated metabolites	*O*-Methylated deaminated metabolites
Amphetamine				
Medulla oblongata	62	350	37	80
Hypothalamus	67	255	25	120
Cortex	85	370	20	110
Desipramine				
Medulla oblongata	60	170	80	115
Hypothalamus	70	75	60	120
Cortex	115	135	60	145
Pheniprazine				
Medulla oblongata	175	850	0	0
Hypothalamus	170	780	0	0
Cortex	170	710	5	45

Unchanged H^3-noradrenaline and metabolites in various regions of the rat brain were measured 1 h after the injection of H^3-noradrenaline into the lateral ventricle. Animals were pretreated with D-amphetamine (15 mg/kg), desipramine (20 mg/kg) or pheniprazine (Catron; 5 mg/kg) 1–2 h before the injection of H^3-NA. Values are means for 5–6 animals and are expressed as percentages of results obtained in normal untreated animals. Note that although amphetamine and desipramine inhibit the accumulation of H^3-NA to approximately the same extent, amphetamine in addition produces large changes in the pattern of metabolites. These changes are similar to those observed after the inhibition of monoamine oxidase with pheniprazine. In addition amphetamine, but not desipramine, is a potent inhibitor of monoamine oxidase *in vitro*.

can cause a depletion of radioactive noradrenaline from brain stores labelled by a previous intraventricular injection of H^3-noradrenaline (Glowinski & Axelrod, 1966). However, the changes in the level of brain noradrenaline do not correlate very closely with the behavioural effects produced by various doses of the drug (Smith, 1965), and the effects of the drug on behaviour are greatly increased in the reserpinized animal, in which brain catecholamine stores are already severely depleted. Recent studies of the metabolism and storage of labelled noradrenaline in various regions of the rat brain after pretreatment with D-amphetamine revealed that the drug has very complex actions on brain catecholamines. In this study we obtained evidence that after amphetamine treatment the metabolism of noradrenaline by monoamine oxidase in the brain was markedly inhibited, thus reviving an old theory that the actions of amphetamine may be due to its ability to act as a competitive inhibitor of monoamine oxidase in the brain (Blaschko, Richter & Schlossman, 1937). In addition D-amphetamine acts as an inhibitor of the uptake of exogenous catecholamines into catecholamine neurones in the brain (Table 10.5). In certain brain areas pretreatment with the drug also appeared to inhibit the conversion of H^3-dopamine to noradrenaline, suggesting yet another possible action for the drug. Amphetamine may thus act in a highly complex manner on brain catecholamines, and it is by no means clear that any one of these actions is the crucial one for explaining the powerful behavioural effects of the drug. As with other drugs which act on behaviour and on brain catecholamines, the demonstration that the drug has a specific action on some particular aspect of brain chemistry does not *per se* prove anything about the mechanism of action of the drug in producing the behavioural change. It will require much more detailed and exhaustive studies to establish which, if any, of the biochemical actions of amphetamine results in central excitation.

5. Concluding remarks: the status and possible functions of central adrenergic mechanisms

Noradrenaline and dopamine as possible candidates for central transmitter substances can now be said to fulfil almost all of the usual criteria (Fuxe *et al.* 1965). The amines are present in high concentrations in the terminals of specific systems of catecholamine-containing neurones in the brain. Enzyme systems for the synthesis and degradation of the catecholamines exist either in or in close proximity to these catecholamine-containing neurones. The presence of an efficient uptake mechanism in the catecholamine-containing neurones may represent a mechanism for the inactivation of the catecholamines after their release from central synapses. There is also evidence, at least in the case of noradrenaline, that the amines are stored in specialized particles or vesicles within the catecholamine nerve terminals.

Microelectrophoretic studies have indicated that many nerve cells exist in the brain which respond to noradrenaline (Bloom, Oliver & Salmoiraghi, 1963; Bradley & Wolstencroft, 1965), the distribution of catecholamine terminals in the brain suggests that at least some of these cells receive an innervation by catecholamine-terminals (Fuxe *et al.* 1965). It remains only to have conclusive evidence that noradrenaline and dopamine are released from catecholamine-containing nerve terminals in response to nerve impulses. A release of noradrenaline from the isolated spinal cord has been demonstrated on stimulation of the anterior end of the cord (Andén, Carlsson, Hillarp & Magnusson, 1965), but this experimental system is so gross that one could wish to have the results of similar experiments on more precisely defined experimental systems. If noradrenaline synthesis is inhibited, stimulation of the bulbospinal catecholamine-containing nuclei resulted in a more rapid disappearance of fluorescent catecholamine-containing terminals in the spinal cord (Dahlström, Fuxe, Kernell & Sedvall, 1965). Prolonged stimulation of the amygdaloid nuclei in cats also results in a disappearance of fluorescent nerve terminals from various brain regions, and an accompanying fall in the noradrenaline content of these regions (Fuxe & Gunne, 1964; Gunne & Reis, 1963; Reis & Gunne, 1965).

The only evidence suggesting a release of dopamine has been the report of McLennan (1964) that the output of dopamine from the caudate nucleus of the cat (recorded by chemical analysis of the perfusate outflow of a 'push-pull' cannula) was increased by electrical stimulation of the nucleus centromedianus.

The results concerning the release of catecholamines in response to nerve impulses, although suggestive, will still require refinement and confirmation before they are generally accepted. Nevertheless, there seems to be good reason to believe that noradrenaline and dopamine both act as neurotransmitter substances in the central nervous system.

The precise functions of the adrenergic neurone systems in the brain and spinal cord, however, remain obscure. The possible role of dopamine in extrapyramidal motor functions and the defects in 'dopaminergic' systems in the Parkinson syndrome have often been discussed (see review by Hornykiewicz, 1966; Sourkes, 1964). The possible role of noradrenaline as an excitatory transmitter in the hypothalamic–limbic forebrain reward system is also intriguing (Poschel & Ninteman, 1963, 1964; Olds, 1962). The intracerebral implantation of crystals of noradrenaline in various brain areas has suggested that noradrenaline may participate in alerting reactions (Yamaguchi, Ling & Marczynski, 1964; Miller, 1964). It seems probable that the ubiquitous distribution of catecholamine-containing neurones in the central nervous system reflects the involvement of these substances as transmitters in a wide variety of different neuronal systems, the details of which will only emerge gradually in coming years.

REFERENCES

Alberici, M., De Lores Arnaiz, G. R. & De Robertis, E. (1965). Catechol-*O*-methyl transferase in nerve endings of rat brain. *Life Sciences* **4**, 1951–1960.

Andén, N. E., Carlsson, A., Dahlström, A., Fuxe, K., Hillarp, N.-Å. & Larsson, K. (1964). Demonstration and mapping out of nigro-neostriatal dopamine neurones. *Life Sciences* **3**, 523–530.

Andén, N. E., Carlsson, A., Hillarp, N.-Å. & Magnusson, T. (1965). Noradrenaline release by nerve stimulation of the spinal cord. *Life Sciences* **4**, 129–132.

Andén, N. E., Häggendal, J., Magnusson, T. & Rosengren, E. (1964). The time course of the disappearance of noradrenaline and 5-hydroxytryptamine in the spinal cord after transection. *Acta physiol. scand.* **62**, 115–118.

Andén, N. E., Magnusson, T. & Rosengren, E. (1965). Occurence of dihydroxyphenylalanine decarboxylase in nerves of the spinal cord and sympathetically innervated organs. *Acta physiol. scand.* **64**, 127–135.

Aprison, M. H. & Takahashi, R. (1965). Biochemistry of the avian central nervous system: II. *J. Neurochem.* **12**, 221–230.

Axelrod, J., Albers, R. W. & Clemente, C. D. (1959). Distribution of catechol-*O*-methyl transferase in the nervous system and other tissues. *J. Neurochem.* **5**, 68–72.

Bagchi, S. P. & McGeer, P. L. (1964). Some properties of tyrosine hydroxylase from the caudate nucleus. *Life Sciences*, **3**, 1195–1200.

Bertler, A. (1961). Occurrence and localization of catecholamines in human brain. *Acta physiol. scand.* **51**, 97–107.

Bertler, A. & Rosengren, E. (1959*a*). Occurrence and distribution of catechol amines in brain. *Acta physiol. scand.* **47**, 350–361.

Bertler, A. & Rosengren, E. (1959*b*). On the distribution in brain of monoamines and of enzymes responsible for their formation. *Experientia* **15**, 382–384.

Blaschko, H., Richter, D. & Schlossman, H. (1937). The oxidation of adrenaline and other amines. *Biochem. J.* **31**, 2187–2196.

Bloom, F. E., Oliver, A. P. & Salmoiraghi, G. C. (1963). The responsiveness of individual hypothalamic neurones to microelectrophoretically administered endogenous amines. *Int. J. Neuropharmac.* **2**, 181–193.

Bogdanski, D. F., Weissbach, H. & Udenfriend, S. (1957). The distribution of serotonin, 5-hydroxytryptophan decarboxylase and monoamine oxidase in brain. *J. Neurochem.* **1**, 272–278.

Bradley, P. B. & Wolstencroft, J. H. (1965). Actions of drugs on single neurones in the brain stem. *Br. med. Bull.* **21**, 15–18.

Brodie, B. B., Bogdanski, D. F. & Bonomi, L. (1964). Formation, storage and metabolism of serotonin (5-hydroxytryptamine) and catecholamines in lower vertebrates. In *Comparative Neurochemistry* (ed. D. Richter), pp. 367–377. Oxford: Pergamon Press.

Brodie, B. B., Spector, S. & Shore, P. A. (1959). Interaction of drugs with norepinephrine in the brain. *Pharmac. Rev.* **11**, 548–564.

Burack, W. R. & Draskoćzy, P. R. (1964). The turnover of endogenously labelled catecholamines in several regions of the sympathetic nervous system. *J. Pharmac. exp. Ther.* **144**, 66–75.

Carlsson, A. (1959). The occurrence, distribution and physiological role of catecholamines in the nervous system. *Pharmac. Rev.* **11**, 490–493.

Carlsson, A. (1964). Functional significance of drug-induced changes in brain monoamine levels. *Prog. Brain Res.* **8**, 9–27.

Carlsson, A., Dahlström, A., Fuxe, K. & Hillarp, N.-Å. (1965). Failure of reserpine to deplete noradrenaline neurones of α-methyl noradrenaline formed from α-methyl-DOPA. *Acta pharmac. toxic.* **22**, 270–276.

Carlsson, A., Falck, B., Fuxe, K. & Hillarp, N.-Å. (1964). Cellular localization of monoamines in the spinal cord. *Acta physiol. scand.* **60**, 112–119.

Carlsson, A., Falck, B. & Hillarp, N.-Å. (1962). Cellular localization of brain monoamines. *Acta physiol. scand.* **56**, Suppl. 196.

Carlsson, A., Lindqvist, M., Magnusson, T. & Waldeck, B. (1958). On the presence of 3-hydroxytyramine in brain. *Science* **127**, 471.

Carlsson, A., Rosengren, E., Bertler, A. & Nilsson, J. (1957). The effect of reserpine on the metabolism of catecholamines. In *Psychotropic Drugs* (ed. Garattini and Ghetti), pp. 363–372. Amsterdam: Elsevier.

Chruściel, T. J. (1960). Observations on the localization of noradrenaline in homogenates of dog's hypothalamus. In *Adrenergic Mechanisms* (ed. Vane, Wolstenholme and O'Connor), p. 539. London: Churchill.

Dahlström, A. & Fuxe, K. (1964*a*). Evidence for the existence of monoamine-containing neurones in the central nervous system. I. Demonstration of monoamines in the cell bodies of brain stem neurons. *Acta physiol scand.* **62**, Suppl. 232.

Dahlström, A. & Fuxe, K. (1964*b*). A method for the demonstration of adrenergic nerve fibres in peripheral nerves. *Z. Zellforsch. mikrosk. Anat.* **62**, 602–607.

Dahlström, A. & Fuxe, K. (1964*c*). A method for the demonstration of adrenergic nerve fibres in the central nervous system. *Acta physiol. scand.* **60**, 293–294.

Dahlström, A. & Fuxe, K. (1965). Evidence for the existence of monoamine neurons in the central nervous system. II. Experimentally induced changes in the intraneuronal amine levels of bulbospinal neuron systems. *Acta physiol. scand.* **64**, Suppl. 247.

Dahlström, A., Fuxe, K. & Hillarp, N.-Å. (1965). Site of action of reserpine. *Acta pharmac. toxic.* **22**, 277–292.

Dahlström, A., Fuxe, K., Kernell, D. & Sedvall, G. (1965). Reduction of monoamine stores in the terminals of bulbospinal neurons following stimulation in the medulla oblongata. *Life Sciences* **4**, 1207–1212.

De Iraldi, A. P., Duggan, F. & De Robertis, E. (1963). Adrenergic synaptic vesicles in the anterior hypothalamus of the rat. *Anat. Rec.* **145**, 521–531.

De Lores Arnaiz, G. R. & De Robertis, E. (1962). Cholinergic and non-cholinergic nerve endings in the rat brain. II. Subcellular localization of monoamine oxidase and succinate dehydrogenase. *J. Neurochem.* **9**, 503–508.

De Robertis, E., De Iraldi, A. P., De Lores Arnaiz, G. R. & Zieher, L. M. (1965). Synaptic vesicles from the rat hypothalamus. Isolation and norepinephrine content. *Life Sciences* **4**, 193–201.

Dengler, H. J., Michaelson, I. A., Spiegel, H. E. & Titus, E. O. (1962). The uptake of labelled norepinephrine by isolated brain and other tissues of the cat. *Int. J. Neuropharmac.* **1**, 23–38.

Dengler, H. J., Spiegel, H. E. & Titus, E. O. (1961). Uptake of tritium labelled norepinephrine in brain and other tissues of cat *in vitro*. *Science, N.Y.* **133**, 1072–1073.

Ehringer, H. & Hornykiewicz, O. (1960). Verteilung von Noradrenalin und Dopamin im Gehirn des Menschen und ihr Verhatten bei Erkrankungen des Extrapyramidelen Systems. *Klin. Wschr.* **24**, 1236–1239.

Everett, G. M. & Weigard, R. G. (1962). Central amines and behavioral states: a critique and new data. *Proc.* 1*st Int. Pharmac. Meeting*, vol. VIII, pp. 85–92. Oxford: Pergamon Press.

Fuxe, K. (1964). Cellular localization of monoamines in the median eminence and infundibular stem of some mammals. *Z. Zellforsch. mikrosk. Anat.* **61**, 710–724.

Fuxe, K. (1965*a*). Evidence for the existence of monoamine-containing neurons in the central nervous system. III. The monoamine nerve terminal. *Z. Zellforsch. mikrosk. Anat.* **65**, 572–596.

Fuxe, K. (1965*b*). Evidence for the existence of monoamine-containing neurons in the central nervous system. IV. Distribution of monoamine nerve terminals in the central nervous system. *Acta physiol. scand.* **64**, Suppl. 247.

Fuxe, K. (1966). Discussion. (Second International Catecholamine Symposium.) *Pharmac. Rev.* **18**, 641.

Fuxe, K., Dahlström, A. & Hillarp, N.-Å. (1965). Central monoamine neurons and monoamine neuro-transmission. *Proc. Int. Un. Physiol. Sciences*, vol. IV, pp. 419–434. XXIIIth Int. Cong. Tokyo. Excerpta Medica Foundation.

Fuxe, K. & Gunne, L. M. (1964). Depletion of the amine stores in brain catecholamine terminals on amygdaloid stimulation. *Acta physiol. scand.* **62**, 493–494.

Fuxe, K. & Hillarp, N.-Å. (1964). Uptake of L-DOPA and noradrenaline by central catecholamine neurons. *Life Sciences* **3**, 1403–1406.

Gaddum, J. H. (1963). Chemical transmission in the central nervous system. *Experientia* **19**, 741.

Gey, K. F. & Pletscher, A. (1964). Effects of chlorpromazine on the metabolism of DL-2-C^{14}-DOPA in the rat. *J. Pharmac. exp. Ther.* **145**, 337–343.

Glowinski, J. & Axelrod, J. (1966). Effects of drugs on the disposition of H^3-norepinephrine in the rat brain. (Second International Catecholamine Symposium.) *Pharmacol. Rev.* **18**, 775–786.

Glowinski, J. & Baldessarini, R. (1966). Brain catecholamines. *Pharmac. Rev.* (In the Press.)

Glowinski, J. & Iversen, L. L. (1966*a*). Regional studies of catecholamines in the rat brain. I. The disposition of H^3-norepinephrine, H^3-dopamine and H^3-dopa in various regions of the brain. *J. Neurochem.* **13**, 655–669.

Glowinski, J. & Iversen, L. L. (1966*b*). Regional studies of catecholamines in the rat brain. III. Subcellular distribution of endogenous and exogenous catecholamines in various brain regions. *Biochem. Pharm.* **15**, 977–987.

Glowinski, J., Iversen, L. L. & Axelrod, J. (1966*a*). Storage and synthesis of norepinephrine in the reserpine-treated rat brain. *J. Pharmac. ex. Ther.* **151**, 385–399.

Glowinski, J., Iversen, L. L. & Axelrod, J. (1966*b*). Regional studies of catecholamines in the rat brain. IV. Effects of drugs on the disposition and metabolism of H^3-norepinephrine and H^3-dopamine. *J. Pharmac. exp. Ther.* **153**, 30–41.

Glowinski, J., Kopin, I. J. & Axelrod, J. (1965). Metabolism of H^3-norepinephrine in the rat brain. *J. Neurochem.* **12**, 25–30.

Glowinski, J., Snyder, S. H. & Axelrod, J. (1966). Subcellular localization of H^3-norepinephrine in the rat brain and the effect of drugs. *J. Pharmac. exp. Ther.* **152**, 282–292.

Gray, E. G. & Whittaker, V. P. (1962). The isolation of nerve endings from brain: an electron miscroscopic study of cell fragments derived by homogenization and centrifugation. *J. Anat.* **96**, 79–88.

Gunne, L. M. & Reise, D. J. (1963). Changes in brain catecholamines associated with electrical stimulation of amygdaloid nucleus. *Life Sciences* **2**, 804–809.

Häggendal, J. & Lindqvist, M. (1964). Disclosure of labile monoamine fractions in brain and their correlation to behaviour. *Acta physiol. scand.* **60**, 351–357.

Hamberger, B. & Masuoka, D. (1965). Localization of catecholamine uptake in rat brain slices. *Acta pharmac. tox.* **22**, 363–368.

Heller, A. & Moore, R. Y. (1965). Effect of central nervous system lesions on brain monoamines in the rat. *J. Pharmac. exp. Ther.* **150**, 1–9.

Holzbauer, M. & Vogt, M. (1956). Depression by reserpine of the noradrenaline concentration in hypothalamus of the cat. *J. Neurochem.* **1**, 8–11.

Hornykiewicz, O. (1966). Dopamine and brain function. *Pharmac. Rev.* **18**, 925–964.

Iversen, L. L. & Glowinski, J. (1966). Regional studies of catecholamines in the rat brain. II. Rate of turnover of catecholamines in various brain regions. *J. Neurochem.* **13**, 671–682.

Kuntzman, R., Shore, P, A., Bogdanski, D. & Brodie, B. B. (1961). Microanalytical procedures for fluorimetric assay of brain DOPA-5HTP decarboxylase, norepinephrine and serotonin and a detailed mapping of decarboxylase activity in brain. *J. Neurochem.* **6**, 226–232.

Laverty, R., Michaelson, I. A., Sharman, D. F. & Whittaker, V. P. (1963). The subcellular localization of dopamine and acetyl choline in the dog caudate nucleus. *Br. J. Pharmac. Chemother.* **21**, 482–490.

Levi, R. & Maynert, E. W. (1964). The subcellular localization of brain stem norepinephrine and 5-hydroxytryptamine in stressed rats. *Biochem. Pharmac.* **13**, 615–621.

McGeer, P. L., Bagchi, S. P. & McGeer, E. G. (1965). Subcellular localization of tyrosine hydroxylase in beef caudate nucleus. *Life Sciences* **4**, 1859–1867.

McGeer, P. L. & McGeer, E. G. (1964). Formation of adrenaline by brain tissue. *Biochem. biophys. Res. Commun.* **17**, 502–507.

McGeer, P. L., Wada, J. A. & McGeer, E. G. (1963). Correlations between central aromatic amine levels and behavioral tests following administration of psychoactive drugs. *Recent Adv. Biol. Psychiat.* **5**, 228–236.

McLennan, H. (1964). The release of acetylcholine and 3-hydroxytyramine from the caudate nucleus, *J. Physiol., Lond.* **174**, 152–161.

Mannarino, E., Kirshner, N. & Nashold, B. S. (1963). The metabolism of C^{14}-noradrenaline by cat brain *in vivo*. *J. Neurochem.* **10**, 373–379.

Matsuoka, M., Ishii, S., Shimizu, N. & Imaizumi, R. (1965). Effect of Win. 18501–2 on the content of catecholamine and the number of catecholamine-containing granules in the rabbit hypothalamus. *Experientia* **21**, 121–125.

Maynert, E. W., Levi, R. & De Lorenzo, A. J. D. (1964). The presence of norepinephrine and 5-hydroxytryptamine in vesicles from disrupted nerve-ending particles. *J. Pharmac. exp. Ther.* **144**, 385–392.

Milhaud, G. & Glowinski, J. (1962). Métabolisme de la dopamine-^{14}C dans le cerveau du rat. Étude du mode d'administration. *C. r. hebd. Séanc. Acad. Sci., Paris* **255**, 203–205.

Milhaud, G. & Glowinski, J. (1963). Métabolisme de la noradrenaline-^{14}C dans le cerveau du rat. *C. r. hebd. Séanc. Acad. Sci., Paris* **256**, 1033–1035.

Miller, N. E. (1964). Chemical coding of behaviour in the brain. *Science, N.Y.* **148**, 328–335.

Montagu, K. A. (1957). Catechol compounds in rat tissues and in brains of different animals. *Nature, Lond.* **180**, 244–245.

Moore, K. E. (1963). Toxicity and catecholamine releasing action of D- and L-amphetamine in isolated and aggregated mice. *J. Pharmac. exp. Ther.* **142**, 6–12.

Olds, J. (1962). Hypothalamic substrates of reward. *Physiol. Rev.* **42**, 554–604.

Poschel, B. P. H. & Ninteman, F. W. (1963). Norepinephrine: a possible excitatory neurohormone of the reward system. *Life Sciences* **2**, 782–788.

Poschel, B. P. H. & Ninteman, F. W. (1964). Excitatory antidepressant effects of monoamine oxidase inhibitors on the reward system of the brain. *Life Sciences* **3**, 903–910.

Potter, L. T. & Axelrod, J. (1963). Subcellular localization of catecholamines in the tissues of the rat. *J. Pharmac. exp. Ther.* **142**, 291–298.

Pscheidt, G. R. & Haber, B. (1965). Regional distribution of dihydroxyphenylalanine and 5-hydroxytryptophan decarboxylase and of biogenic amines in the chicken central nervous system. *J. Neurochem.* **12**, 613–618.

Pscheidt, G. R., Morpurgo, C. & Himwich, H. E. (1964). Studies on norepinephrine and 5-hydroxytryptamine in various species. In *Comparative Neurochemistry* (ed. D. Richter), pp. 401–412. Oxford: Pergamon Press.

Reise, D. J. & Gunne, L. M. (1965). Brain catecholamines: relation to the defense reaction evoked by amygdaloid stimulation in cat. *Science, N.Y.* **149**, 450–451.

Reivitch, M. & Glowinski, J. (1966). (In preparation.)

Roos, B. E. & Werdinius, B. (1962). Effect of reserpine on level of 5-hydroxyindoleacetic acid in brain. *Life Sciences* **3**, 105–107.

Ross, S. B. & Renyi, A. L. (1964). Blocking action of sympathomimetic amines on the uptake of tritiated noradrenaline by mouse cerebral cortex tissues *in vitro*. *Acta pharmac. tox.* **21**, 226–239.

Salmoiraghi, G. C., Costa, E. & Bloom, F. E. (1965). Pharmacology of central synapses. *A. Rev. Pharmac.* **5**, 213–234.

Sano, I., Gamo, T., Kakimoto, Y., Tamiguchi, K., Takesada, M. & Nishinuma, K. (1959). Distribution of catechol compounds in human brain. *Biochim. biophys. Acta* **32**, 586–587.

Smith, C. B. (1965). Effects of D-amphetamine upon brain amine content and locomotor activity of mice. *J. Pharmac. exp. Ther.* **147**, 96–102.

Snyder, S. H., Glowinski, J. & Axelrod, J. (1965). The storage of norepinephrine and some of its derivatives in brain synaptosomes. *Life Sciences* **4**, 797–807.

Sourkes, T. L. (1964). Cerebral and other diseases with disturbance of amine metabolism. *Prog. Brain Res.* **8**, 186–200.

Stein, L. (1964). Self stimulation of the brain and the central stimulant action of amphetamine. *Fedn. Proc.* **23**, 836–850.

Udenfriend, S. & Creveling, C. R. (1959). Localization of dopamine-β-oxidase in brain. *J. Neurochem.* **4**, 350–352.

Udenfriend, S. & Zaltzman-Nirenberg, P. (1963). Norepinephrine and 3,4-dihydroxy-phenylethylamine turnover in guinea pig brain *in vivo*. *Science* **142**, 394–396.

Vogt, M. (1954). The concentration of sympathin in different parts of the central nervous system under normal conditions and after the administration of drugs. *J. Physiol., Lond.* **123**, 451–481.

Vogt, M. (1957). Sympathomimetic amines in the central nervous system. *Br. med. Bull.* **13**, 166–171.

Vogt, M. (1964). Some species differences in the responses of animals to drugs. In *Comparative Neurochemistry* (ed. D. Richter), pp. 395–399. Oxford: Pergamon Press.

Weil-Malherbe, H., Axelrod, J. & Whitby, L. G. (1961). The blood–brain barrier for catecholamines in different regions of the brain. In *Regional Neurochemistry* (ed. S. Kety and J. Elkes), pp. 234–292. Oxford: Pergamon Press.

Whittaker, V. P. (1965). The application of subcellular fractionation techniques to the study of brain function. *Prog. Biophys. molec. Biol.* **15**, 39–96.

Yamaguchi, N., Ling, G. M., & Marczynski, T. J. (1964). The effects of chemical stimulation of the preoptic region, nucleus centralis medialis or brain stem reticular formation with regard to sleep and wakefulness. *Recent Adv. Biol. Psychiat.* **6**, 9.

INDEX